Stress Analysis by the Finite Element Method for Practicing Engineers

Stress Analysis by the Finite Element Method for Practicing Engineers

William H. Bowes
Carleton University

Leslie T. Russell
Nova Scotia Technical College

Lexington Books
D.C. Heath and Company
Lexington, Massachusetts
Toronto London

Library of Congress Cataloging in Publications Data

Bowes, William H

Stress analysis by the finite element method for practicing engineers.

Bibliography: p.

Includes index.

1. Structural design. 2. Finite element method. I. Russell, L. T. II. Title.

TA658.2.B68 624'.1771 75-6279

ISBN 0-669-99903-2

Published simultaneously in Canada

Printed in the United States of America

International Standard Book Number: 0-669-99903-2

Library of Congress Catalog Card Number: 75-6279

Contents

List of Figures

List of Tables

Preface

Computers and programs in Finite Element Methods have now reached a level of development which makes the method a practical tool for design engineers. Every design engineer should have for his use a set of programs that will determine stresses in shapes approximating his designs. The use of such programs is essential in designing more refined and reliable products. If North American designs are to continue to meet competition from abroad, such new techniques can not be ignored.

It is to present the Finite Element Method as a working tool for design engineers that the authors have written this book. Many more scholarly and complete works are in existence, but engineers find these difficult to understand through self-study. Even when they are understood the engineer must still write his own programs which may, in itself, require more time and effort than he can afford to expend. The present work is an attempt to explain the subject with sufficient clarity so that an engineer can master the fundamentals of the subject by self-study. An extensive background is not necessary; merely a knowledge of Fortran and a familiarity with basic matrix operations.

Advanced elements (which are being developed at an accelerating rate) are not dealt with, as they are well explained in many other books and journals. The engineer who has need for such elements will be better able to understand the papers written about them after he is thoroughly acquainted with the simple elements. He will also be able to adapt the programs presented in this text to use the advanced elements.

In the text, seven main programs are presented. In each case the theory is developed, flow charts are shown, sample programs are given and the Fortran statements are listed. All subroutines are listed in the Appendix. This makes it possible for the reader to punch a complete set of working programs for the solution of certain basic stress problems. Starting with these programs and their subroutines the designer can write programs to solve other categories of problems that are of particular interest to him. This can be done with a minimum of effort by calling upon the subroutines that are presented.

The text deals with the subject in such a manner that it can readily be understood by practicing engineers and can be effectively used for a first course at the graduate level. Advanced undergraduate students should also be able to benefit from studying this presentation of the Finite Element Method.

Stress Analysis by the Finite Element Method for Practicing Engineers

1 The Finite Element Method

The Finite Element Method has only recently become a useful tool for the analysis of stress. The first complete presentation of the method was made in 1956 by Turner, Clough, Martin, and Topp [14]. Williams [15] in 1954, had come very close to discovering the Finite Element Method, but his method still relied heavily on the finite difference approximations. In 1941, Hrennikoff [6] presented a useful but limited method of solving plane stress problems by replacing plate elements with bars thus reducing a plate problem to one in structural analysis. The Finite Element Method would not be practical without the availability of fast computers with large amounts of storage space, and the development of methods and machines have progressed simultaneously. Problems which a decade ago would have baffled the specialist in stress analysis, or at best been solved by very crude approximations, are now solved with ease in a routine manner by simply describing the problem with appropriate data cards and processing the data by standard finite element computer programs.

The Finite Element Method was first used in aircraft design where the great rewards for slight improvements in weight made it profitable to pioneer the development and application of this new technique. The next natural area for application of this method was in ship design, and now there is no reason why all design engineers should not make use of the method, thus greatly improving the reliability of their products.

Exact stress analysis by the theory of elasticity can be done in a very few bodies of simple geometry and those only when the loads are of particular form. Finite difference approximations have extended the categories of problems that can be solved by enabling the problems to be converted from those expressed in partial differential equations to simultaneous linear equations that can be solved to give approximate solutions. Although this extends the realm of tractable problems, setting up each new problem is a formidable programming task.

The Finite Element Method does not deal with the partial differential equations of equilibrium or compatibility. Rather it converts the problem directly into one requiring the solution of simultaneous linear equations. Solving problems by the Finite Element Method therefore requires a working knowledge of matrix operations and the reader is referred to Chapter 2 of Desai and Abel [3] or Appendix 1 in Zienkiewicz [16] for this material. Using finite elements, the types of problems that can be solved are almost limitless. Once the programming is done for any one category of problem, all problems of that class can be analyzed without further programming. In classical treatments, very different techniques are required to solve problems in different categories such as plane stress, shells, and plates in bending. For all of these and many other categories of problems, the techniques of the Finite Element Method are remarkably similar. They are also remarkably easy to develop and once programmed can be used by unskilled assistants.

The universal nature of the method makes it possible to solve with equal ease the stress problems of civil and of mechanical engineering. Since problems of a structural type are usually easier to visualize, they will be used to introduce the principles that have wider application. Thus we will be dealing with a pin-jointed truss to establish the basic principles. When this has been done the step to shells of revolution or to plane stress is found to be relatively easy.

A statically determinate structure, Figure 1-1a, can be analyzed easily by elementary methods of statics to determine the stresses in the members. If redundant members such as *BC* and *BE* are added, Figure 1-1b, the problem becomes more difficult; and more advanced methods are necessary to obtain a solution. The stiffness method could be used or it could be solved by the Finite Element Method. For this type of problem, the two methods would be almost identical and we will solve this truss by the Finite Element Method, not because another method of solution is required but rather to develop the method in the context of a problem that is rather simple and hence avoid having the method obscured by the complexity of the problem. The elements into which the structure should be broken are rather obvious, viz, the members *AC*, *CE*, etc. In other problems, the choice of elements will require some skill on the part of the analyst.

The steps in the solution to this problem and to any other by the Finite Element Method are, in the broadest of terms:

i. Break the "structure" into elements.
ii. Find the stiffness of each element.
iii. Assemble the element stiffnesses into a matrix which defines the stiffness of the whole assembly.
iv. For the given load system, use the stiffness matrix to determine the displacements of the nodes (pins in this pin-jointed truss case).
v. Using the displacements, calculate the stresses in the members.

Stiffness of a Pin-Ended Bar

When dealing with springs, it is common practice to specify the spring

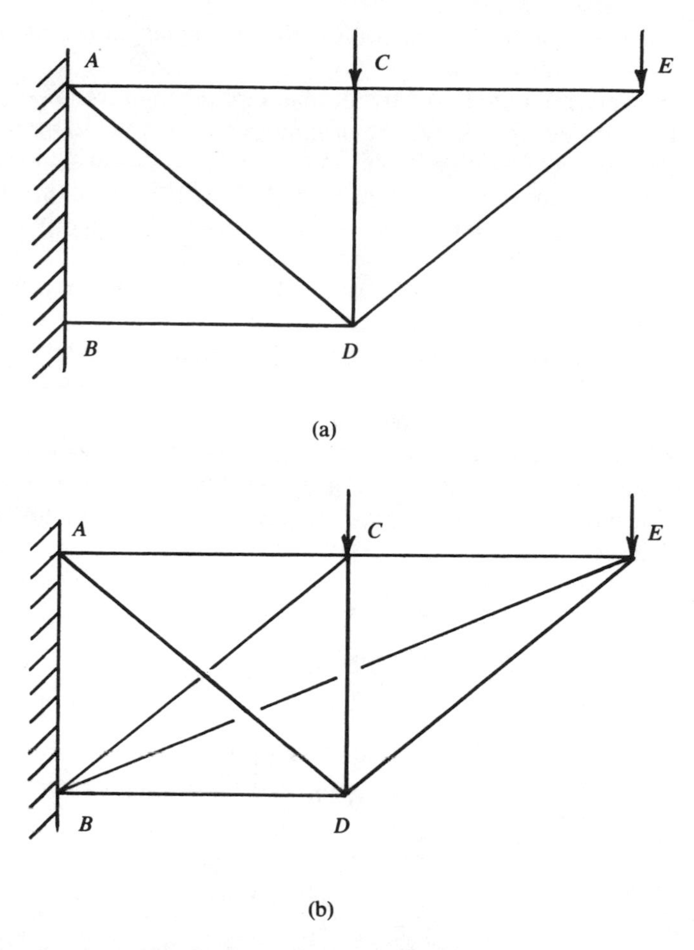

Figure 1-1. Structural Problems

modulus, k, in force per unit elongation, which gives the total description of the spring stiffness. It is then easy to determine the force in the spring for any change in length from

$$F = k \Delta L$$

For a structural member, k can be easily calculated but would be of little use since in a typical structural element such as DE of Figure 1-1 the forces that must be considered have four components, which are dependent upon four components of displacement. Then the stiffness of the structural element must be expressed as a 4×4 stiffness matrix, K, which relates displacement and force components by equation (1.1),

$$\mathbf{f} = \mathsf{K}\boldsymbol{\delta} \tag{1.1}$$

where **f** and $\boldsymbol{\delta}$ are column vectors of forces and displacements as shown in Figure 1-2.

The ends are indicated now by numbers rather than letters, for convenience. The broken line shows the original, unloaded position of the bar. The δ components show the displacements that occur during loading and the f components are the forces necessary to hold the bar in this displaced position. It is assumed that the displacements are sufficiently small that changes in θ during the displacement are insignificant.

There are simple methods which give the values of the sixteen elements of K. However, they are applicable to this case only and cannot be extended to more advanced cases, hence will not be presented here. The method that follows is that which must be used to deal with all more advanced cases and this is the justification for introducing it at this stage.

We will start by finding the stiffness of the element with a different set of axes and components, as shown in Figure 1-3. The first objective will be to establish formulas that give u' and v' displacements at any point P on the element. These functions will be assumed to be

$$u' = \alpha_1 + \alpha_2 x'$$
$$v' = \alpha_3 + \alpha_4 x'$$

which can be written

$$\begin{Bmatrix} u' \\ v' \end{Bmatrix} = \begin{bmatrix} 1 & x' & 0 & 0 \\ 0 & 0 & 1 & x' \end{bmatrix} \begin{Bmatrix} \alpha_1 \\ \alpha_2 \\ \alpha_3 \\ \alpha_4 \end{Bmatrix} \tag{1.2}$$

where the left-hand member is a function vector containing the two functions describing the displacements in the x' and y' directions.

In (1.2) there are four unknown coefficients $\alpha_1, \alpha_s, \alpha_3, \alpha_4$; but we know four displacement components at the ends of the element and hence can evaluate these coefficients. This gives

$$u'_{(x'=x'_1)} = \delta'_1 = [1 \quad x'_1 \quad 0 \quad 0]\boldsymbol{\alpha}$$
$$u'_{(x'=x'_2)} = \delta'_2 = [1 \quad x'_2 \quad 0 \quad 0]\boldsymbol{\alpha}$$
$$v'_{(x'=x'_1)} = \delta'_3 = [0 \quad 0 \quad 1 \quad x'_1]\boldsymbol{\alpha}$$
$$v'_{(x'=x'_2)} = \delta'_4 = [0 \quad 0 \quad 1 \quad x'_2]\boldsymbol{\alpha}$$

which can be written

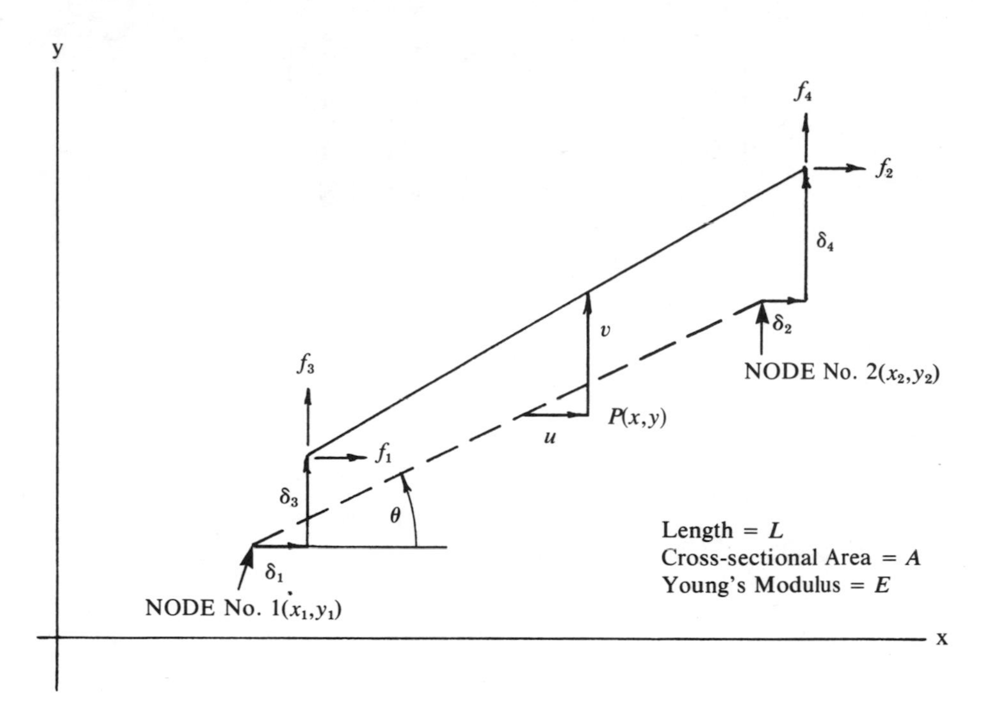

Figure 1-2. Displacement of Structural Element

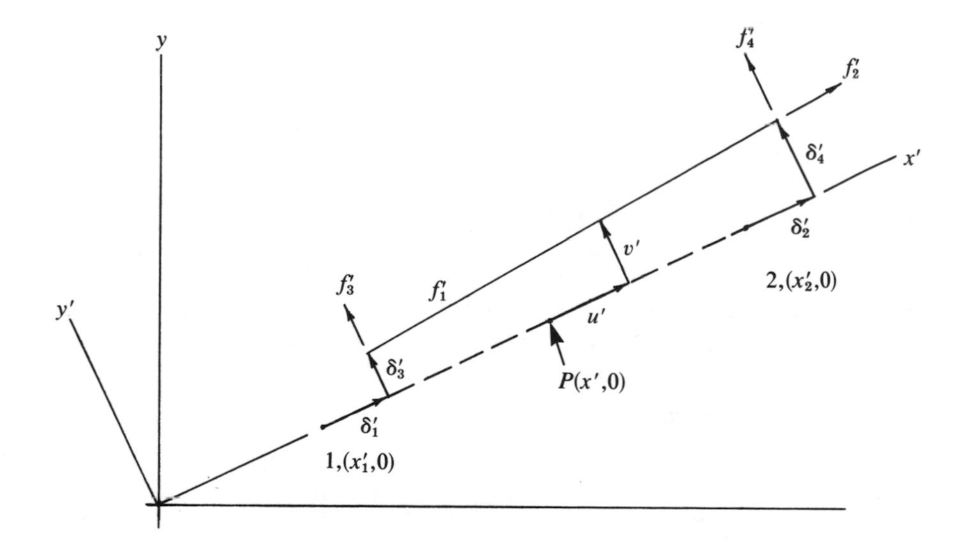

Figure 1-3. Structural Element Rotated Axes

$$\begin{Bmatrix} \delta'_1 \\ \delta'_2 \\ \delta'_3 \\ \delta'_4 \end{Bmatrix} = \begin{bmatrix} 1 & x'_1 & 0 & 0 \\ 1 & x'_2 & 0 & 0 \\ 0 & 0 & 1 & x'_1 \\ 0 & 0 & 1 & x'_2 \end{bmatrix} \begin{Bmatrix} \alpha_1 \\ \alpha_2 \\ \alpha_3 \\ \alpha_4 \end{Bmatrix}$$

Or,

$$\boldsymbol{\delta}' = \mathsf{A}\boldsymbol{\alpha}$$

where

$$\boldsymbol{\alpha} = \begin{Bmatrix} \alpha_1 \\ \alpha_2 \\ \alpha_3 \\ \alpha_4 \end{Bmatrix} \quad \text{and} \quad \mathsf{A} = \begin{bmatrix} 1 & x'_1 & 0 & 0 \\ 1 & x'_2 & 0 & 0 \\ 0 & 0 & 1 & x'_1 \\ 0 & 0 & 1 & x'_2 \end{bmatrix}$$

From which

$$\boldsymbol{\alpha} = \mathsf{A}^{-1}\boldsymbol{\delta}' \tag{1.3}$$

The inversion of A could be done directly but it can more readily be inverted as two half-size matrices as follows:

$$\mathsf{A} = \begin{bmatrix} \overline{\mathsf{A}} & 0 \\ 0 & \overline{\mathsf{A}} \end{bmatrix} \quad \text{where} \quad \overline{\mathsf{A}} = \begin{bmatrix} 1 & x'_1 \\ 1 & x'_2 \end{bmatrix}$$

Then,

$$\mathsf{A}^{-1} = \begin{bmatrix} \overline{\mathsf{A}}^{-1} & 0 \\ 0 & \overline{\mathsf{A}}^{-1} \end{bmatrix}$$

The advantage gained in inverting the smaller matrices in this case is not particularly great, but later when A is large, commonly 12×12, the advantage will be quite significant.

In the present case,

$$\overline{\mathsf{A}}^{-1} = \frac{1}{x'_2 - x'_1} \begin{bmatrix} x'_2 & -x'_1 \\ -1 & 1 \end{bmatrix} = \frac{1}{L} \begin{bmatrix} x'_2 & -x'_1 \\ -1 & 1 \end{bmatrix}$$

and,

and,

$$\mathsf{A}^{-1} = \frac{1}{L}\begin{bmatrix} x_2' & -x_1' & 0 & 0 \\ -1 & 1 & 0 & 0 \\ 0 & 0 & x_2' & -x_1' \\ 0 & 0 & -1 & 1 \end{bmatrix} \tag{1.4}$$

Substituting, (1.3) into (1.2), we have

$$\begin{Bmatrix} u' \\ v' \end{Bmatrix} = \begin{bmatrix} 1 & x' & 0 & 0 \\ 0 & 0 & 1 & x' \end{bmatrix} \mathsf{A}^{-1}\boldsymbol{\delta}' \tag{1.5}$$

where A^{-1} is given by (1.4); but the substitution will not be made at this stage.

From elementary strength of materials, the axial strain is given by

$$\varepsilon = \frac{\partial u'}{\partial x'} = \begin{bmatrix} \dfrac{\partial}{\partial x'} & 0 \end{bmatrix}\begin{Bmatrix} u' \\ v' \end{Bmatrix}$$

and, using (1.5),

$$\begin{aligned} \varepsilon &= \begin{bmatrix} \dfrac{\partial}{\partial x'} & 0 \end{bmatrix} \begin{bmatrix} 1 & x' & 0 & 0 \\ 0 & 0 & 1 & x' \end{bmatrix} \mathsf{A}^{-1}\boldsymbol{\delta}' \\ &= [0 \;\; 1 \;\; 0 \;\; 0]\mathsf{A}^{-1}\boldsymbol{\delta}' \\ &= \mathsf{B}\mathsf{A}^{-1}\boldsymbol{\delta}' \end{aligned} \tag{1.6}$$

where $\mathsf{B} = [0 \;\; 1 \;\; 0 \;\; 0]$, a very simple matrix in this case. In more advanced cases, B will be quite complicated.

Provided the material obeys Hooke's Law, stress can be expressed as $\sigma = E\varepsilon$ where all quantities are scalars. Later, for example, when stress in a plane is considered, both stress and strain will have three components and these combined components can be treated as vectors. They are then related through a 3×3 matrix made up of the elastic constants which can be written

$$\boldsymbol{\sigma} = \mathsf{D}\boldsymbol{\varepsilon} \tag{1.7}$$

In order to make the equations take on the form of those that will be appearing later, we will use these matrix notations, remembering that in the present case each matrix in (1.7) has a single element and is really a scalar quantity. In keeping with this, we should rewrite (1.6) as

$$\boldsymbol{\varepsilon} = \mathsf{BA}^{-1}\boldsymbol{\delta}' \qquad (1.8)$$

Consider the element in Figure 1-3 which has been displaced from its zero-stress state by amounts given in $\boldsymbol{\delta}'$ (considered here as known components) and held in this position by forces $\mathbf{f}'$.

In this state, the stresses are given by

$$\boldsymbol{\sigma} = \mathsf{DBA}^{-1}\boldsymbol{\delta}' \qquad (1.9)$$

which comes from substituting (1.8) into (1.7).

From this state imagine that we give the element an additional virtual displacement [12] $\boldsymbol{\delta}'^*$. This will put an additional strain into the fibres, which is given by

$$\boldsymbol{\varepsilon}^* = \mathsf{BA}^{-1}\boldsymbol{\delta}'^* \qquad (1.10)$$

During the virtual displacement, the force components of $\mathbf{f}'$ move through distances given in $\boldsymbol{\delta}'^*$ and do an amount of work, w_E, which is given by

$$w_E = \delta_1'^* f_1' + \delta_2'^* f_2' + \delta_3'^* f_3' + \delta_4'^* f_4' = \boldsymbol{\delta}'^{*T}\mathbf{f}'$$

At the same time, work is done by the existing stress in the fibres as the virtual strain is applied. This change in internal work, w_I, is given by

$$w_I = \int_{\text{vol}} \boldsymbol{\varepsilon}^{*T}\boldsymbol{\sigma}\, dv$$

When (1.9) and (1.10) are substituted,

$$w_I = \int_{\text{vol}} [\mathsf{BA}^{-1}\boldsymbol{\delta}'^*]^T\mathsf{DBA}^{-1}\boldsymbol{\delta}'\, dv$$

$$= \int_{\text{vol}} \boldsymbol{\delta}'^{*T}[\mathsf{A}^{-1}]^T\mathsf{B}^T\mathsf{DBA}^{-1}\boldsymbol{\delta}'\, dv$$

Since all terms to be integrated are, in this case, constants

$$w_I = \boldsymbol{\delta}'^{*T}[\mathsf{A}^{-1}]^T\mathsf{B}^T\mathsf{DBA}^{-1}\boldsymbol{\delta}' \times \text{vol}$$

But the external work done during the virtual displacement must equal the change in internal strain energy, therefore

$$w_E = w_I$$

or

$$\boldsymbol{\delta}'^{*T}\mathbf{f}' = \boldsymbol{\delta}'^{*T}[A^{-1}]^T\mathsf{B}^T\mathsf{DBA}^{-1}\boldsymbol{\delta}'\,\text{vol}$$

hence

$$\mathbf{f}' = \text{vol}[\mathsf{A}^{-1}]^T\mathsf{B}^T\mathsf{DBA}^{-1}\boldsymbol{\delta}' \qquad (1.11)$$

When (1.11) is compared with (1.1), it is evident that, for the case being considered,

$$\mathsf{K}' = \mathrm{vol}[\mathsf{A}^{-1}]^T\mathsf{B}^T\mathsf{D}\mathsf{B}\mathsf{A}^{-1} \tag{1.12}$$

Equation (1.12) gives the stiffness of the element and relates displacement and force components parallel to the rotated axes of Figure 1-3. The analysis of the complete structure will be made with reference to the original axis system and hence the element stiffness must be transformed so that it relates the displacement and force components in the original axis system.

In Figure 1-4, the vector quantity P, which subsequently will be force and later displacement, has components $P\cos\phi$ in the x direction and $P\sin\phi$ in the y direction. Based on this, the force components in the original x and y directions can be written as

$$f_1 = f_1' \cos\theta + f_3' \cos(\pi/2 + \theta) = f_1' \cos\theta - f_3' \sin\theta$$

$$f_2 = f_2' \cos\theta + f_4' \cos(\pi/2 + \theta) = f_2' \cos\theta - f_4' \sin\theta$$

$$f_3 = f_1' \sin\theta + f_3' \sin(\pi/2 + \theta) = f_1' \sin\theta + f_3' \cos\theta$$

$$f_4 = f_2' \sin\theta + f_4' \sin(\pi/2 + \theta) = f_2' \sin\theta + f_4' \cos\theta$$

which in matrix format becomes

$$\mathbf{f} = \begin{bmatrix} \cos\theta & 0 & -\sin\theta & 0 \\ 0 & \cos\theta & 0 & -\sin\theta \\ \sin\theta & 0 & \cos\theta & 0 \\ 0 & \sin\theta & 0 & \cos\theta \end{bmatrix} \mathbf{f}'$$

$$\mathbf{f} = \mathsf{T}\mathbf{f}'$$

Similarly:

$$\boldsymbol{\delta} = \mathsf{T}\boldsymbol{\delta}'$$

Hence,

$$\mathbf{f}' = \mathsf{T}^{-1}\mathbf{f} \tag{1.13}$$

and

$$\boldsymbol{\delta}' = \mathsf{T}^{-1}\boldsymbol{\delta} \tag{1.14}$$

Inverting T, we find that

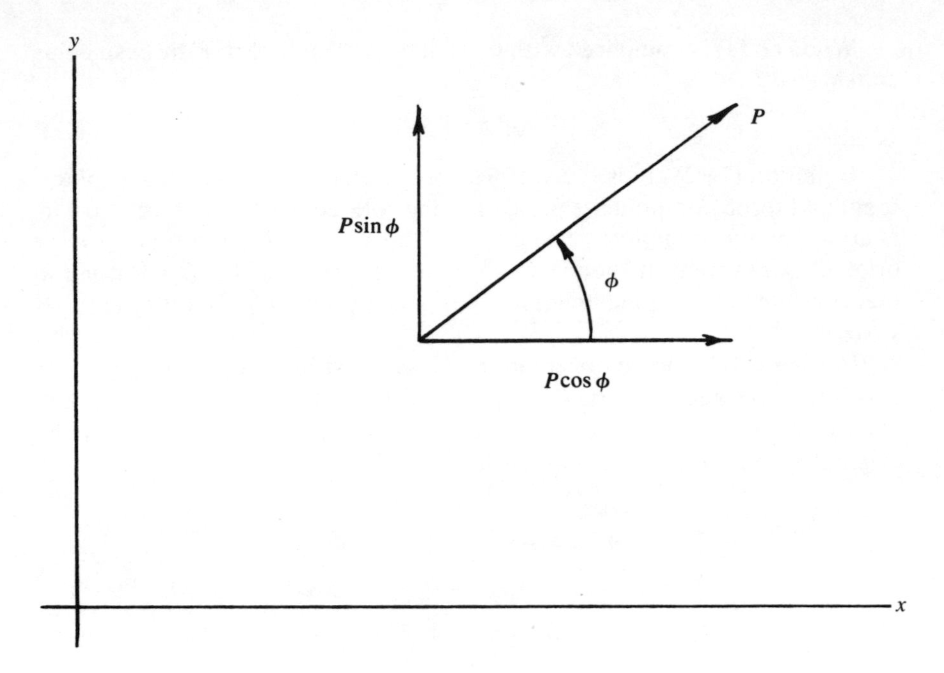

Figure 1-4. Vector Diagram

$$\mathsf{T}^{-1} = \begin{bmatrix} \cos\theta & 0 & \sin\theta & 0 \\ 0 & \cos\theta & 0 & \sin\theta \\ -\sin\theta & 0 & \cos\theta & 0 \\ 0 & -\sin\theta & 0 & \cos\theta \end{bmatrix} \tag{1.15}$$

From which it may be seen that

$$\mathsf{T}^{-1} = \mathsf{T}^{T} \tag{1.16}$$

Substituting (1.13) and (1.14) into (1.11), we find that

$$\mathsf{T}^{-1}\mathbf{f} = \text{vol}[\mathsf{A}^{-1}]^{T}\mathsf{B}^{T}\mathsf{DBA}^{-1}\mathsf{T}^{-1}\boldsymbol{\delta}$$

Hence,

$$\mathbf{f} = \text{vol}\,\mathsf{T}[\mathsf{A}^{-1}]^{T}\mathsf{B}^{T}\mathsf{DBA}^{-1}\mathsf{T}^{T}\boldsymbol{\delta}$$

and

$$\mathsf{K} = \text{vol}\,\mathsf{T}[\mathsf{A}^{-1}]^{T}\mathsf{B}^{T}\mathsf{DBA}^{-1}\mathsf{T}^{T}$$

But all values on the right side are known in terms of the dimensions of the element, substituting these we get

K =

$$L \times A \times \begin{bmatrix} \cos\theta & 0 & -\sin\theta & 0 \\ 0 & \cos\theta & 0 & -\sin\theta \\ \sin\theta & 0 & \cos\theta & 0 \\ 0 & \sin\theta & 0 & \cos\theta \end{bmatrix} \frac{1}{L} \begin{bmatrix} x_2' & -1 & 0 & 0 \\ -x_1' & 1 & 0 & 0 \\ 0 & 0 & x_2' & -1 \\ 0 & 0 & -x_1' & 1 \end{bmatrix} \begin{Bmatrix} 0 \\ 1 \\ 0 \\ 0 \end{Bmatrix}$$

$$E[0\ 1\ 0\ 0]\ \frac{1}{L} \begin{bmatrix} x_2' & -x_1' & 0 & 0 \\ -1 & 1 & 0 & 0 \\ 0 & 0 & x_2' & -x_1' \\ 0 & 0 & -1 & 1 \end{bmatrix} \begin{bmatrix} \cos\theta & 0 & \sin\theta & 0 \\ 0 & \cos\theta & 0 & \sin\theta \\ -\sin\theta & 0 & \cos\theta & 0 \\ 0 & -\sin\theta & 0 & \cos\theta \end{bmatrix}$$

$$= \frac{A \times E}{L} \begin{bmatrix} \cos^2\theta & -\cos^2\theta & \sin\theta\cos\theta & -\sin\theta\cos\theta \\ -\cos^2\theta & \cos^2\theta & -\sin\theta\cos\theta & \sin\theta\cos\theta \\ \sin\theta\cos\theta & -\sin\theta\cos\theta & \sin^2\theta & -\sin^2\theta \\ -\sin\theta\cos\theta & \sin\theta\cos\theta & -\sin^2\theta & \sin^2\theta \end{bmatrix} \tag{1.17}$$

Substituting

$$\cos\theta = \frac{x_2 - x_1}{L} \quad \text{and} \quad \sin\theta = \frac{y_2 - y_1}{L}$$

in (1.17), and letting $X = x_2 - x_1$ and $Y = y_2 - y_1$, gives

$$K = \frac{AE}{L^3} \begin{bmatrix} X^2 & -X^2 & XY & -XY \\ -X^2 & X^2 & -XY & XY \\ XY & -XY & Y^2 & -Y^2 \\ -XY & XY & -Y^2 & Y^2 \end{bmatrix} \tag{1.18}$$

The General Stiffness Matrix

Having established the stiffness of the pin-ended structural element it is now necessary to compile a matrix which gives the stiffness of a complete structure such as that shown in Figure 1-5. Every pinned joint, henceforth

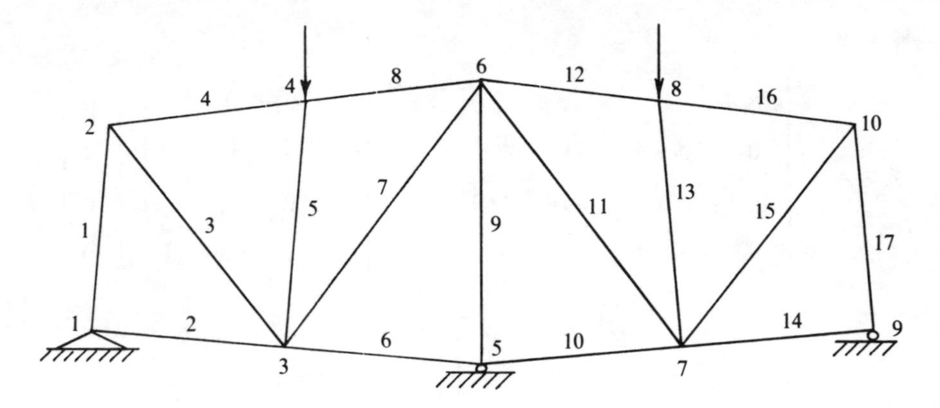

Figure 1-5. Truss

referred to as a node, is considered to have two degrees of freedom and hence two components of displacement which are to be determined. The supports dictate that some displacement components are zero but this is not taken into account until a later stage. Now if there are N nodes the structure has $2N$ components of displacement and $2N$ components of force acting on the structure, and many of these are zero. The displacement vector, **U**, containing all displacement components, and the corresponding load vector, **F**, each have twenty components in the case shown in Figure 1-5. Consequently, in

$$\mathbf{F} = \mathsf{K}_A \mathbf{U} \tag{1.19}$$

K_A is a 20×20 matrix.

Some systematic method must be established for relating the **F** and **U** components to the nodes and the following system will be used in this work. The horizontal force component at node M is in position $2M - 1$ in the force vector and the vertical component is in position $2M$. Components of displacement are similarly located.

Each structural member contributes to certain of the elements of the general stiffness matrix and by totalling these contributions from all members the complete general stiffness matrix is obtained. To appreciate how this is done, it is best to start by imagining that all members are infinitely flexible, that is, they offer no resistance to deformation nor will the structure. Hence, K_A is filled with zeros. Now assume that one member, 3-6, has its proper rigidity and that its stiffness matrix K has been determined, that is all K_{ij} are known in (1.1), which is rewritten as

$$\begin{Bmatrix} f_1 \\ f_2 \\ f_3 \\ f_4 \end{Bmatrix} = \begin{bmatrix} K_{11} & K_{12} & K_{13} & K_{14} \\ K_{21} & K_{22} & K_{23} & K_{24} \\ K_{31} & K_{32} & K_{33} & K_{34} \\ K_{41} & K_{42} & K_{43} & K_{44} \end{bmatrix} \begin{Bmatrix} \delta_1 \\ \delta_2 \\ \delta_3 \\ \delta_4 \end{Bmatrix}$$

For the particular element being considered, δ_1 is the displacement component corresponding to the 5th component in the general displacement vector $\mathbf{U}$.

That is,

$$\begin{array}{lll} \delta_1 \equiv U_5 & & f_1 \equiv F_5 \\ \delta_2 \equiv U_{11} & \text{and} & f_2 \equiv F_{11} \\ \delta_3 \equiv U_6 & & f_3 \equiv F_6 \\ \delta_4 \equiv U_{12} & & f_4 \equiv F_{12} \\ \text{a.} & & \text{b.} \end{array} \qquad (1.20)$$

This means that the entries appearing in column 1 of K will appear in column 5 of the general stiffness matrix; those of column 2 will appear in column 11; etc.

The element force component f_1 corresponds to the general force component F_5, hence the first row of K will appear in the 5th row of K_A, etc. When all elements of K have been added, the general stiffness matrix will be all zeros except for the 16 entries which are the contributions to the stiffness from member 3-6; and K_A gives the stiffness of the structure which has most of its members still in the infinitely flexible condition. At this stage, K_A would be as shown in Figure 1-6. We next give another member its correct rigidity and find its stiffness matrix, K, by (1.18). Then determine, as above, from the numbers of the nodes at the ends of the member the positions in K_A where the elements of K make a contribution and add these elements to any values that may already be in these locations. By repeating this process for all members of the assembly, each and every member makes its proper contribution to the general stiffness matrix which is then totally known.

If we were to generate K_A for the case under consideration and inspect the values of the 400 elements, we would observe three features:

i. Some of the values would be negative and others positive but we would find only positive values on the principal diagonal. The law of conservation of energy dictates that these be positive.

ii. There is symmetry about the principal diagonal.

$$K_A = \begin{bmatrix}
0 & 0 & 0 & 0 & 0 & 0 & 0 & 0 & 0 & 0 & 0 & 0 & 0 & 0 & 0 & 0 & 0 & 0 & 0 & 0 \\
0 & 0 & 0 & 0 & 0 & 0 & 0 & 0 & 0 & 0 & 0 & 0 & 0 & 0 & 0 & 0 & 0 & 0 & 0 & 0 \\
0 & 0 & 0 & 0 & 0 & 0 & 0 & 0 & 0 & 0 & 0 & 0 & 0 & 0 & 0 & 0 & 0 & 0 & 0 & 0 \\
0 & 0 & 0 & 0 & 0 & 0 & 0 & 0 & 0 & 0 & 0 & 0 & 0 & 0 & 0 & 0 & 0 & 0 & 0 & 0 \\
0 & 0 & 0 & 0 & K_{11} & K_{13} & 0 & 0 & 0 & 0 & K_{12} & K_{14} & 0 & 0 & 0 & 0 & 0 & 0 & 0 & 0 \\
0 & 0 & 0 & 0 & K_{31} & K_{33} & 0 & 0 & 0 & 0 & K_{32} & K_{34} & 0 & 0 & 0 & 0 & 0 & 0 & 0 & 0 \\
0 & 0 & 0 & 0 & 0 & 0 & 0 & 0 & 0 & 0 & 0 & 0 & 0 & 0 & 0 & 0 & 0 & 0 & 0 & 0 \\
0 & 0 & 0 & 0 & 0 & 0 & 0 & 0 & 0 & 0 & 0 & 0 & 0 & 0 & 0 & 0 & 0 & 0 & 0 & 0 \\
0 & 0 & 0 & 0 & 0 & 0 & 0 & 0 & 0 & 0 & 0 & 0 & 0 & 0 & 0 & 0 & 0 & 0 & 0 & 0 \\
0 & 0 & 0 & 0 & 0 & 0 & 0 & 0 & 0 & 0 & 0 & 0 & 0 & 0 & 0 & 0 & 0 & 0 & 0 & 0 \\
0 & 0 & 0 & 0 & K_{21} & K_{23} & 0 & 0 & 0 & 0 & K_{22} & K_{24} & 0 & 0 & 0 & 0 & 0 & 0 & 0 & 0 \\
0 & 0 & 0 & 0 & K_{41} & K_{43} & 0 & 0 & 0 & 0 & K_{42} & K_{44} & 0 & 0 & 0 & 0 & 0 & 0 & 0 & 0 \\
0 & 0 & 0 & 0 & 0 & 0 & 0 & 0 & 0 & 0 & 0 & 0 & 0 & 0 & 0 & 0 & 0 & 0 & 0 & 0 \\
0 & 0 & 0 & 0 & 0 & 0 & 0 & 0 & 0 & 0 & 0 & 0 & 0 & 0 & 0 & 0 & 0 & 0 & 0 & 0 \\
0 & 0 & 0 & 0 & 0 & 0 & 0 & 0 & 0 & 0 & 0 & 0 & 0 & 0 & 0 & 0 & 0 & 0 & 0 & 0 \\
0 & 0 & 0 & 0 & 0 & 0 & 0 & 0 & 0 & 0 & 0 & 0 & 0 & 0 & 0 & 0 & 0 & 0 & 0 & 0 \\
0 & 0 & 0 & 0 & 0 & 0 & 0 & 0 & 0 & 0 & 0 & 0 & 0 & 0 & 0 & 0 & 0 & 0 & 0 & 0 \\
0 & 0 & 0 & 0 & 0 & 0 & 0 & 0 & 0 & 0 & 0 & 0 & 0 & 0 & 0 & 0 & 0 & 0 & 0 & 0 \\
0 & 0 & 0 & 0 & 0 & 0 & 0 & 0 & 0 & 0 & 0 & 0 & 0 & 0 & 0 & 0 & 0 & 0 & 0 & 0 \\
0 & 0 & 0 & 0 & 0 & 0 & 0 & 0 & 0 & 0 & 0 & 0 & 0 & 0 & 0 & 0 & 0 & 0 & 0 & 0
\end{bmatrix}$$

Figure 1-6. Partially Completed Stiffness Matrix

iii. A large region in the lower left corner, and another in the upper right, is filled with zeros.

Later, use will be made of ii and iii to conserve memory space by not storing the redundant elements or the zeros.

Finding the Unknown Displacements

At this point, all elements of K_A are known, as are a few of **U** and most of **F**. In the case shown in Figure 1-5, the supports require that $U_1 = 0$, $U_2 = 0$, $U_{10} = 0$, and $U_{18} = 0$; while in the force vector **F**, F_1, F_2, F_{10}, and F_{18} are the only components that are not known. This correspondence between knowns and unknowns in **F** and **U** always exists. The equation $\mathbf{F} = \mathsf{K}_A\mathbf{U}$ expresses in matrix form a system of twenty simultaneous equations with twenty unknowns but this differs from the classical simultaneous equation problem in that the unknowns are not all the **U** vector. This problem can be solved by an elimination process which will not be developed here as it leads to many complications. The practical method of solution is that of Payne and Irons [10], which uses ficticious components of force in place of the unknowns, makes some changes in K_A, and solves as though all the **U** components are unknown. Let us say that U_i is a known component. Then a ficticious component is put into the force vector given by

$$F_i = cK_{Aii}U_i$$

where c is a large number. (In practice, $c = 10^{12}$ is satisfactory.)

The stiffness matrix is changed by substituting cK_{Aii} for K_{Aii}. The ith equation is now treated in the same way as all the others, the changes being such that they ensure that the equations when solved will give the original known value for U_i. By treating each row that contains a known U component in this way, the problem is changed to one in which all the **F** components have numerical values and all the **U** components are treated as unknowns. Solving this classical problem will give all values of **U**. A simple substitution of the known **U** into (1.19) using the original unmodified K_A gives all **F** components, some of which have not been known up to this point and others, that were known, are recalculated. A comparison of these components with the original known components serves as a check on the process.

Finding the Stresses in the Members

Each member is now considered in turn and its $\boldsymbol{\delta}$ components determined

$$
K_A = 10^6 \begin{bmatrix}
.986 & 0 & -.0267 & -.160 & -.960 & .160 & 0 & 0 & 0 \\
0 & .986 & -.160 & -.960 & .160 & -.0267 & 0 & 0 & 0 \\
-.0267 & -.160 & 1.22 & -.00998 & -.236 & .330 & -.960 & -.160 & 0 \\
-.160 & -.960 & -.00998 & 1.45 & .330 & -.462 & -.160 & -.0267 & 0 \\
-.960 & .160 & -.236 & .330 & 2.34 & -.302 & -.0114 & -.0916 & -.960 \\
.160 & -.0267 & .330 & -.462 & -.302 & 1.63 & -.0916 & -.733 & .160 \\
0 & 0 & -.960 & -.160 & -.0114 & -.0916 & 2.10 & .487 & 0 \\
0 & 0 & -.160 & -.0267 & -.0916 & -.733 & .478 & .805 & 0 \\
0 & 0 & 0 & 0 & -.960 & .160 & 0 & 0 & 1.92 \\
0 & 0 & 0 & 0 & .160 & -.0267 & 0 & 0 & 0 \\
0 & 0 & 0 & 0 & -.171 & -.256 & -1.13 & -.226 & 0 \\
0 & 0 & 0 & 0 & -.256 & -.384 & -.226 & -.0453 & 0 \\
0 & 0 & 0 & 0 & 0 & 0 & 0 & 0 & -.96 \\
0 & 0 & 0 & 0 & 0 & 0 & 0 & 0 & -.160 \\
0 & 0 & 0 & 0 & 0 & 0 & 0 & 0 & 0 \\
0 & 0 & 0 & 0 & 0 & 0 & 0 & 0 & 0 \\
0 & 0 & 0 & 0 & 0 & 0 & 0 & 0 & 0 \\
0 & 0 & 0 & 0 & 0 & 0 & 0 & 0 & 0 \\
0 & 0 & 0 & 0 & 0 & 0 & 0 & 0 & 0 \\
0 & 0 & 0 & 0 & 0 & 0 & 0 & 0 & 0
\end{bmatrix}
$$

Figure 1-7. Form of Stiffness Matrix (Banded)

by selecting the appropriate components from the **U** vector. By substituting (1.16) into (1.14) and then into (1.9) we have the expression for stress

$$\boldsymbol{\sigma} = \mathsf{DBA}^{-1}\mathsf{T}^T\boldsymbol{\delta} \tag{1.21}$$

For each member, all values on the right side of (1.21) are known and, hence, the stress can be calculated. As there are no variables in any of the matrices in (1.21), $\boldsymbol{\sigma}$ does not vary from point to point within the element. This is obviously as it should be for the type of structure being analyzed. It will be found later when dealing with more advanced elements that many values in B are variables. This will make the integration that must be performed in determining K more difficult and will result in stresses that vary throughout the element.

Stiffness Matrix Band-Width

If the stiffness matrix K_A is assembled for the structure in Figure 1-5, it will resemble that shown in Figure 1-7. Inspecting the matrix reveals that

0	0	0	0	0	0	0	0	0	0	0
0	0	0	0	0	0	0	0	0	0	0
0	0	0	0	0	0	0	0	0	0	0
0	0	0	0	0	0	0	0	0	0	0
.160	−.171	−.256	0	0	0	0	0	0	0	0
−.0267	−.256	−.384	0	0	0	0	0	0	0	0
0	−1.13	−.226	0	0	0	0	0	0	0	0
0	−.226	−.0453	0	0	0	0	0	0	0	0
0	0	0	−.96	−.160	0	0	0	0	0	0
.653	0	−.60	−.16	−.0267	0	0	0	0	0	0
0	2.60	0	−.171	.256	−1.13	.226	0	0	0	0
−.60	0	1.46	.256	−.384	.226	−.0453	0	0	0	0
−.16	−.171	.256	2.34	.302	−.0114	.0916	−.960	−.160	−.236	−.330
−.0267	.256	−.384	.302	1.63	.0916	−.733	−.160	−.0267	−.330	−.462
0	−1.13	.226	−.0114	.0916	2.10	−.478	0	0	−.960	.160
0	.226	−.0453	.0916	−.733	−.487	.805	0	0	.160	−.0267
0	0	0	−.960	−.160	0	0	.986	0	−.0267	.160
0	0	0	−.160	−.0267	0	0	0	.986	.160	−.960
0	0	0	−.236	−.330	−.960	.160	−.0267	.160	1.22	.0099
0	0	0	−.330	−.462	.160	−.026	.160	−.960	.0099	1.45

outside of the two broken lines all elements are zero and also that the whole matrix is symmetrical with respect to the principal diagonal. Hence, all the useful information is contained in the elements on the diagonal and on one side, say to the left and below, to the broken line. Thus, all the significant information can be stored in a much smaller matrix, S, as shown in Figure 1-8. We will retain K_A for convenience in analysis but store the data in S for computations, thus making a considerable saving in the amount of computer memory space required. Also, by doing our matrix operations on S, many unproductive mathematical operations on the zeros that appear in K_A are avoided. There are still some zeros in S but the effort required to avoid them is not worthwhile.

The length of the longest column of elements from the diagonal to the last nonzero entry, referred to here as NBAND, or the band width, must now be determined. To do this let us reconsider member number 7 which joins node 3 to node 6 in Figure 1-5. From Figure 1-6 it is evident that, considering this member only, the band width would be 8, this is because contributions are made to rows that range from a low of 5 to a high of 12. Rows 5 to 12, inclusive, span 8 rows. We would have reached the same

$$
\mathsf{S} = 10^6 \begin{bmatrix}
.986 & .986 & 1.22 & 1.45 & 2.34 & 1.63 & 2.10 & .805 & 1.92 \\
0 & -.160 & -.0099 & .330 & -.302 & -.0916 & .478 & 0 & 0 \\
-.0267 & -.960 & -.236 & -.462 & -.0114 & -.733 & 0 & 0 & 0 \\
-.160 & .160 & .330 & -.160 & -.0916 & .160 & 0 & -.226 & 0 \\
-.960 & -.0267 & -.960 & -.0267 & -.960 & -.0267 & -1.13 & -.0453 & -.96 \\
.160 & 0 & -.160 & 0 & .160 & -.256 & -.226 & 0 & -.160 \\
0 & 0 & 0 & 0 & -.171 & -.384 & 0 & 0 & 0 \\
0 & 0 & 0 & 0 & -.256 & 0 & 0 & 0 & 0
\end{bmatrix}
$$

Figure 1-8. Form of Stiffness Matrix (Essential Data)

conclusion by inspecting the subscripts on the right side of the equivalences in (1.20). If the member connected node L to node M, the range would be from $2L - 1$ to $2M$ and the band width would be $2M - (2L - 1) + 1$ or $2((M - L) + 1)$. Each element may span a different number of rows and the maximum span will determine the actual band width, $NBAND$. As each element is considered, if the difference between the end node numbers is recorded, the maximum difference, $NDIFF$, can be selected and the band width calculated by

$$NBAND = 2(NDIFF + 1)$$

Applying this process to Figure 1-5, it is found that for the given structure $NBAND = 8$.

The numbering system, which is purely arbitrary, has an influence on the band width. If, in Figure 1-5, the node now numbered 2 had been numbered 10, the band width would have been 20, which would have left no corner region containing the zeros. In actual fact, the same numerical values would appear in K_A; but the zero elements would have been mixed with the nonzeros, which would mean more quantities to store and many meaningless operations on zeros. By a skillful assignment of node numbers the band width can be minimized and an inspection of Figure 1-5 shows that no improvement can be made here by changing the numbering. To minimize band width, a simple procedure is to number across the small dimension at one extremity of the structure and then in succeeding adjacent rows until the whole structure has been covered. This scheme has been used in Figure 1-5.

Since the algebra for substituting S into the stiffness equation has not been developed, and would probably be very complicated, we will continue to discuss the equations containing K_A in their original form. When calculations are performed using K_A, each numerical value will be extracted as required from the corresponding location in S.

On page 15 it was stated that after the alterations prescribed by Payne

$$
\left.\begin{matrix}
.653 & 2.60 & 1.46 & 2.34 & 1.63 & 2.10 & .805 & .986 & .986 & 1.22 & 1.45 \\
0 & 0 & .256 & .302 & .0916 & -.478 & 0 & 0 & .160 & .0099 & 0 \\
-.60 & -.171 & -.384 & -.0114 & -.733 & 0 & 0 & -.0267 & -.960 & 0 & 0 \\
-.16 & .256 & .226 & .0916 & -.160 & 0 & .160 & .167 & 0 & 0 & 0 \\
-.0267 & -1.13 & -.0453 & -.960 & -.0267 & -.960 & -.0267 & 0 & 0 & 0 & 0 \\
0 & .226 & 0 & -.160 & -.330 & .160 & 0 & 0 & 0 & 0 & 0 \\
0 & 0 & 0 & -.236 & -.462 & 0 & 0 & 0 & 0 & 0 & 0 \\
0 & 0 & 0 & -.330 & 0 & 0 & 0 & 0 & 0 & 0 & 0
\end{matrix}\right]
$$

and Irons, all **F** components had numerical values and so a classical solution of

$$\mathbf{F} = \mathsf{K}_A\mathbf{U}$$

would give all components of **U**. This might suggest that we would invert K_A to solve for **U**, and indeed this could be done. However, part of the advantage of the banded nature of K_A would then be lost, as its inverse would not be banded and hence a much larger storage space would be required for the inverse of the stiffness matrix than for the essential elements of the matrix itself. The inverse would be necessary if a general solution to the system of equations was required, but normally a solution for a particular set of force components is required. Hence, using the added storage space cannot be justified. Under these conditions, the best method to use is Gauss' elimination scheme with back substitution [7].

A Program to Solve Pin-Jointed Trusses

We will next consider a program, referred to as PT10B, that solves for stresses in the members of a truss under any conceivable system of loads no matter what the degree of redundancy. The purpose in dealing with PT10B is not to present this program as an end in itself but rather to present the methods that have wider application than just this problem. Many of the subroutines are used repeatedly by programs that solve more advanced problems.

The following will be more readily understood if, while reading, reference is repeatedly made to Figure 1-9 PT10B Flow Chart; Figure 1-10, PT10B Fortran Statements; Figure 1-13, Instructions for PT10B Data Deck Preparation; and Figure 1-14, Typical Output for Case Processed by PT10B.

When PT10B execution begins, the first output is a general title which

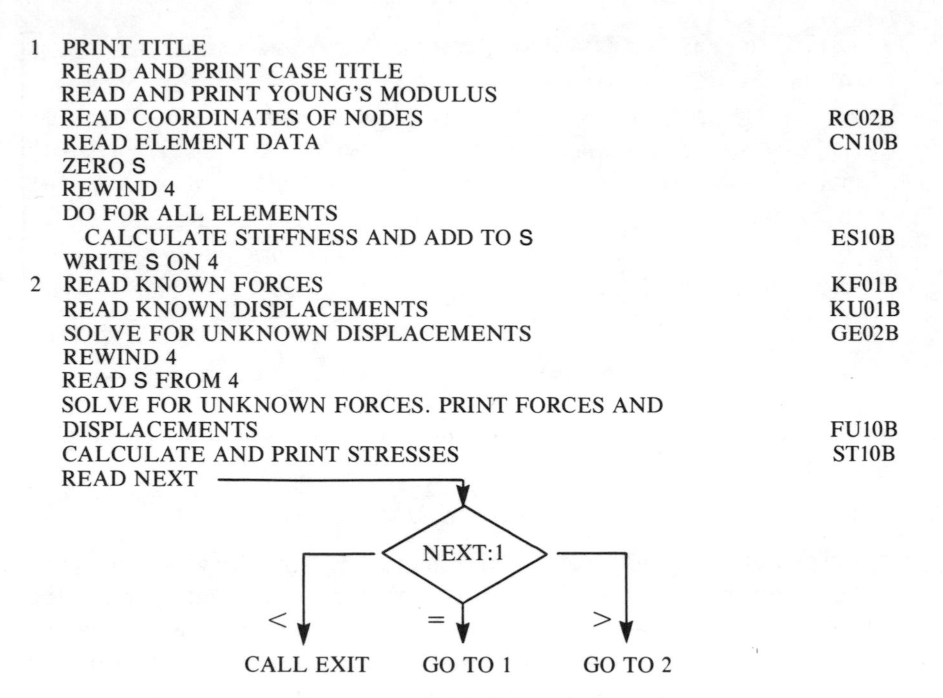

Figure 1-9. PT10B Flow Chart

identifies the program. This is followed by a subtitle which is composed by the user and serves to identify the case being run. The subtitle is punched into the first card of the data deck.

The next data card must contain the modulus of elasticity of the structural material. Cases in which the modulus of elasticity is not the same for all members can not be treated by this program.

In order to define the dimensions of the structure, for computer processing, the user establishes coordinate axes with origin at any convenient location. The axes may be rotated if necessary but must remain at right angles to one another. One data card is prepared for each node, containing the node number and its x and y coordinates. The node numbers must be positive and run sequentially beginning at 1. The computer will read as many node-data cards as there are in the deck and will only terminate this phase and proceed to the next when a zero is read for a node number. Hence in the deck the end of the node-data cards is indicated by a blank card.

The definition of the structure must now be completed by providing the size and location of the members (henceforth referred to as elements). This is done by preparing one data card for each element, containing the element

```
C        *****     MAIN PT10B        *****          OCT 14 1971
C          PIN-JOINTED TRUSS          W H BONES
      DIMENSION S(20,120),U(120),F(120),IFX(120),XY(2,60)
      DIMENSION ARB(100),NCB(2,100)
      DIMENSION NAME(20)
      LBAND=20
    1 WRITE(3,100)
      READ(2,102) NAME
      WRITE(3,103) NAME
      READ(2,106) YMB
      WRITE(3,107) YMB
      CALL RC02B(2,XY,NDF)
      NBAND=0
      CALL CN10B(NCB,ARB,NB,NBAND)
      IF(NBAND-LBAND)5,5,4
    4 WRITE(3,104)
      CALL EXIT
    5 DO 6 I=1,NBAND
      DO 6 J=1,NDF
    6 S(I,J)=0.
      REWIND 4
      DO 8 IB=1,NB
    8 CALL ES10B(IB,NCB,XY,ARB,YMB,S,LBAND)
      DO 10 I=1,NBAND
   10 WRITE(4)(S(I,J),J=1,NDF)
      END FILE 4
    2 CALL KF01B(F,NDF)
      CALL KU01B(U,IFX,NDF)
      CALL GE02B(F,U,S,IFX,NBAND,NDF,LBAND)
      REWIND 4
      DO 20 I=1,NBAND
   20 READ(4)(S(I,J),J=1,NDF)
      CALL FU10B(S,U,NBAND,NDF,2,LBAND)
      CALL ST10B(NB,NCB,XY,U,YMB)
      READ(2,108) NEXT
      IF(NEXT-1)24,1,2
   24 CALL EXIT
  100 FORMAT('1           MAIN PT10B    OCT 14 1971',/,'STRESS IN A PIN-JOI
     1NTED TRUSS')
  102 FORMAT(20A4)
  103 FORMAT('0CASE TITLE   ---   ',20A4)
  104 FORMAT('0BAND WIDTH TOO LARGE')
  106 FORMAT(F10.5)
  107 FORMAT('0YOUNGS MODULUS =',E10.3)
  108 FORMAT(I5)
      END
```

Figure 1-10. PT10B Fortran Statements

number, its cross-sectional area, and the node numbers at its ends. As with the node numbers, the element numbers must be positive and run sequentially beginning at 1.

At this stage, the data read in is sufficient for the program to proceed. It determines the band width and checks to see if that exceeds the limit imposed by certain *DIMENSION* specifications. If the limit is exceeded a message is printed and execution terminated.

The stiffness of each element, in turn, is calculated and accumulated in the proper locations of the S matrix. Since a later operation will alter S and subsequently the original S will be required, at this stage it is preserved by storing on tape or disc.

Loads that are imposed on the structure, other than the reactions, are now required. These are presented as x and y components of loads that are applied at panel points (nodes). Since bending in the members is not taken into consideration, loads at points other than nodes can not be treated. The loads are broken into x and y components and signs associated with the numerical values are used to indicate the direction in the standard way. Each component is numbered as described on page 12.

Load-data cards are prepared by punching one card for each nonzero load component. A load-data card contains the component number and the magnitude of the load component. Any number of load cards may be placed in the data deck with the end of these cards indicated by a blank card.

Advantage should be taken of the fact that only nonzero components need to be read in as data. The program was arranged this way in order to substantially reduce the input data deck in cases where a large number of nodes are not loaded.

The support configuration is defined by reading known displacement components. Pivoted supports and roller supports can thus be represented by reading displacement components equal to zero. If there is settlement at any support this can be treated by reading the amount of the settlement as the value of the appropriate displacement component. Unlike the force components, zero displacement components must be read.

In the data deck, one card must be punched for each known displacement component containing the component number (which is determined in the same way as the force component number) and the value of the displacement component.

As each known displacement component is read a record of its component number is kept by entering 1 in the position in the column vector **IFX** corresponding to the component number. As **IFX** was initially filled with zeros, after all displacement components have been read it is filled with 0's and 1's and serves to indicate which **U** components are known or fixed. Thus **IFX** provides vital information for the method of Payne and Irons. The displacement-data cards are followed by a blank card.

A subroutine, GE02B, takes the data provided at this stage and proceeds to find the unknown displacement components. It uses the method of Payne and Irons, Gauss' elimination scheme, and back substitution to find all the components of **U**. In this process the original values in **S** become written over and hence lost. So the original **S** is recovered by reading it from tape or disc where it was stored earlier in the process. Now by a multiplication equivalent to that in (1.19) all **F** components are evaluated. Many of these were already known; but others, the reactions, are evaluated for the first time. These operations are performed in subroutine FU10B which also prints out all force and displacement components. The reevaluated force components can be compared with the original values as a check.

Taking each element in turn, subroutine ST10B selects from **U** the

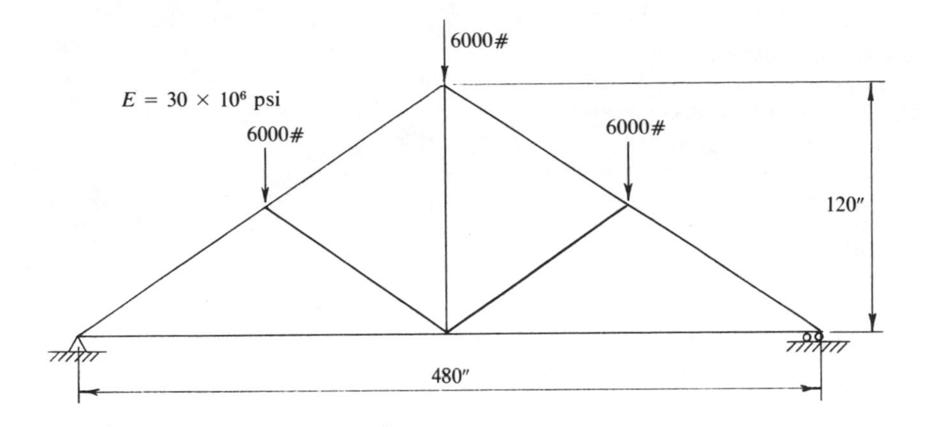

Figure 1-11. King Post Truss

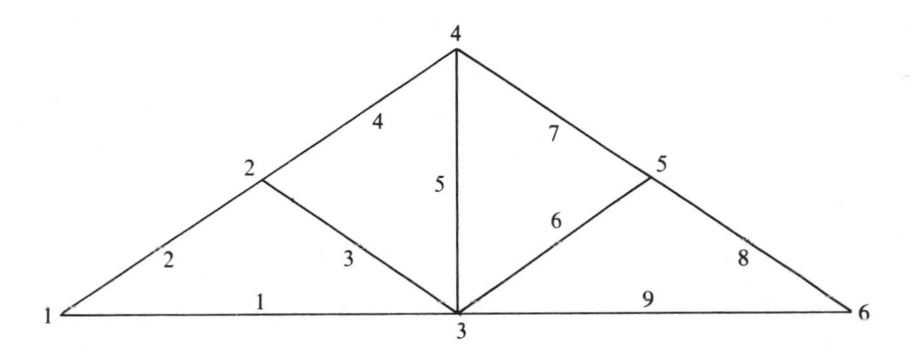

Figure 1-12. Numerical Identification System for Nodes and Elements

components that apply to the element and by the equivalent of equation (1.21) determines and prints the stress.

This completes one solution to the case being considered. If the user wants to study the effect of a different loading pattern or a different support system, the program can be rerun from the point where known load components are read. The user also has the option of executing the program from the beginning with a complete new data deck. The option that the user elects is indicated by a number in another data card.

Example

The King Post truss shown in Figure 1-11 will be used as an example.

Data Deck:

One card containing case title.

One card containing Young's Modulus. Format: F10.5

One card for each node, containing;

Node number, x coord, y coord. Format I5, 2F10.5

One blank card to indicate end of node data.

One card for each element, containing;

Element Number, x section area, Node numbers at ends of element, Format: I5, F10.5, 2I5

One blank card to indicate end of element data.

One card for each known, nonzero force component, containing;

Component number, Force. Format:I5, F10.5

One blank card to indicate end of force data.

One card for each known displacement component, containing;

Component number, Displacement. Format:I5, F10.5

One blank card to indicate end of displacement data.

One card containing NEXT. Format: I5

NEXT = 0; End of job.

NEXT = 1; Execute a new case. Follow by data cards prepared according to all above instructions.

NEXT = 2; Repeat the case just completed but with a new set of known loads and displacements. Follow by data cards described above starting with force data cards.

Figure 1-13. Instructions for PT10B Data Deck Preparation

Problem

Determine the force and displacement components at each panel point, the stress in each member, and the reaction forces. Compare the member stresses as found by the Finite Element Method with the theoretical member stresses.

Solution. To solve any problem by the Finite Element Method the stressed body must first be subdivided into elements. In this problem, since we have only the bar element to use, it is obvious that each truss member should be an element and the pin joint a node. A convenient numerical identification system for nodes and elements is shown in Figure 1-12. The information consisting of dimensions, loads, and boundary conditions is put into the computer according to the instructions in Figure 1-13 and is printed out together with the solution as shown in Figure 1-14.

Comments on Solution

The element developed in this chapter and used in the above problem, the

```
         MAIN PT10B     OCT 14 1971
STRESS IN A PIN-JOINTED TRUSS

CASE TITLE  ---        STRESSES IN KING POST TRUSS

YOUNGS MODULUS =   .300E+08

NODE NO.     X-COORD        Y-COORD
   1       .0000E+00      .0000E+00
   2       .1200E+03      .6000E+02
   3       .2400E+03      .0000E+00
   4       .2400E+03      .1200E+03
   5       .3600E+03      .6000E+02
   6       .4800E+03      .0000E+00

BAR NUMBER     X-SECT AREA     CONNECTS NODES NO.
   1              1.00            1   3
   2              1.20            1   2
   3               .40            2   3
   4               .80            2   4
   5               .30            3   4
   6               .40            3   5
   7               .80            4   5
   8              1.20            5   6
   9              1.00            3   6

BAND WIDTH =      8

  KNOWN NON-ZERO LOADS
COMPONENT NUMBER     LOAD
      4           -.6000E+04
      8           -.6000E+04
     10           -.6000E+04

      KNOWN DISPLACEMENTS
COMPONENT NUMBER    DISPLACEMENT
           1          .0000E+00
           2          .0000E+00
          12          .0000E+00

NODE NO.        FORCE AND DISPLACEMENT COMPONENTS
     1      .3001E-10   .9000E+04  -.1144E-27  -.1677E-12

     2     -.1910E-10  -.6000E+04   .2479E+00  -.6634E+00
     3     -.7276E-11  -.4547E-11   .1440E+00  -.7034E+00
     4     -.1182E-10  -.6000E+04   .1440E+00  -.6234E+00
     5      .2638E-10  -.6000E+04   .4015E-01  -.6634E+00
     6     -.8185E-11   .9000E+04   .2880E+00  -.1677E-12

BAR NO.     STRESS PSI
   1         .1800E+05
   2        -.1677E+05
   3        -.1677E+05
   4        -.1677E+05
   5         .2000E+05
   6        -.1677E+05
   7        -.1677E+05
   8        -.1677E+05
   9         .1800E+05
*EXIT*
```

Figure 1-14. Typical Output for Case Processed by PT10B

pin-ended bar element, because of the assumed displacement function, can represent exactly a linear axial displacement. If the members are of constant cross-section and have only axial loads, the resulting strain is constant and hence the displacement is linear. Because the assumed displace-

Table 1-1
Comparison of Finite Element and Theoretical Solutions

	Stresses (psi)	
Member	*Finite Element Method*	*Theoretical Value*
1	18000	18,000
2	−16770	−16,770
3	−16770	−16,770
4	−16770	−16,770
5	20000	20,000
6	−16770	−16,770
7	−16770	−16,770
8	−16770	−16,770
9	18000	18,000

ment and the actual displacement of the pin-ended bar element are both linear, the finite element solution should yield, except for computer round-off, the exact solution. A comparison of the computer results and the exact answers (see Table 1-1) shows that the finite element results are exact in this case.

2 Triangular Elements

Displacement Functions for Triangular Elements

The ideas introduced in the first chapter will now be extended so that in-plane loaded plate problems can be solved for displacements and stresses. When the assembly was composed of pin-ended members, there was no decision to be made in breaking the structure into elements; each member was treated as an element. A plate that is to be broken into triangles has no natural subdivision and part of the art of analyzing for stresses by the Finite Element Method consists of choosing the elements so as to produce the best results. As before, the elements will be assumed to be connected at the nodes, which will be located at the corners of the triangles; and only loads that are applied at the nodes will be considered. When other loads exist, they will be replaced by equivalent nodal forces.

Consider an elastic plate that is cut into triangles which are then connected together at the nodes only, as in Figure 2-1a. If a load is applied as in Figure 2-1b, it would be expected that the deformations might give shapes such as those shown. That is, gaps might open up along the edges and more deformation than in the original uncut plate would be expected. This must not be allowed, as the flexibility of the system is influenced by such behaviour. By connecting the triangles at the corners, we ensure that the displacements of all triangles joining at any node are the same at that one point. This is expressed by saying that the displacements are compatible at the nodes, which is the case in the original plate. However, at neighboring points on an edge, the displacements in Figure 2-1b are not necessarily the same. If the plate is to behave as it did before it was partitioned into triangles, the compatibility of displacements along the edges, as well as at the nodes, must be assured. This problem did not arise in the pin-jointed structure and we will deal with it later.

Consider a triangle such as that of Figure 2-2. The triangle has been given displacements $\boldsymbol{\delta}$ and the calculation of the force components in $\mathbf{f}$ is desired. This could be done if the stiffness matrix, K, were known by

$$\mathbf{f} = \mathsf{K}\boldsymbol{\delta} \tag{2.1}$$

Therefore, determination of the 36 values in K is necessary. Begin by assuming displacement functions u and v that would give the horizontal and vertical displacements at any point $p(x, y)$ in the triangle. Invariably, in the

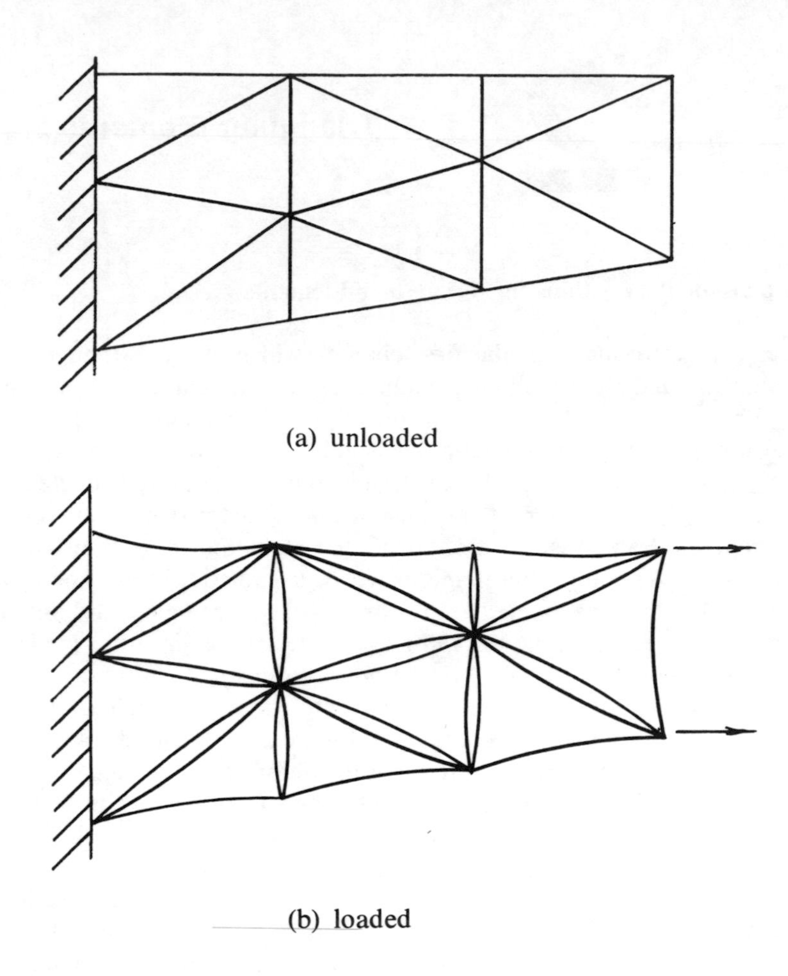

(a) unloaded

(b) loaded

Figure 2-1. Triangular Subdivision of Elastic Plate

Finite Element Method the displacement functions are taken as polynomials. Since the horizontal displacements are known at only three points, that is the nodes, no more than three coefficients in the polynomial can be included and hence the highest-order expression that we should attempt is

$$u = \alpha_1 + \alpha_2 x + \alpha_3 y$$

Similarly choose

$$v = \alpha_4 + \alpha_5 x + \alpha_6 y$$

or,

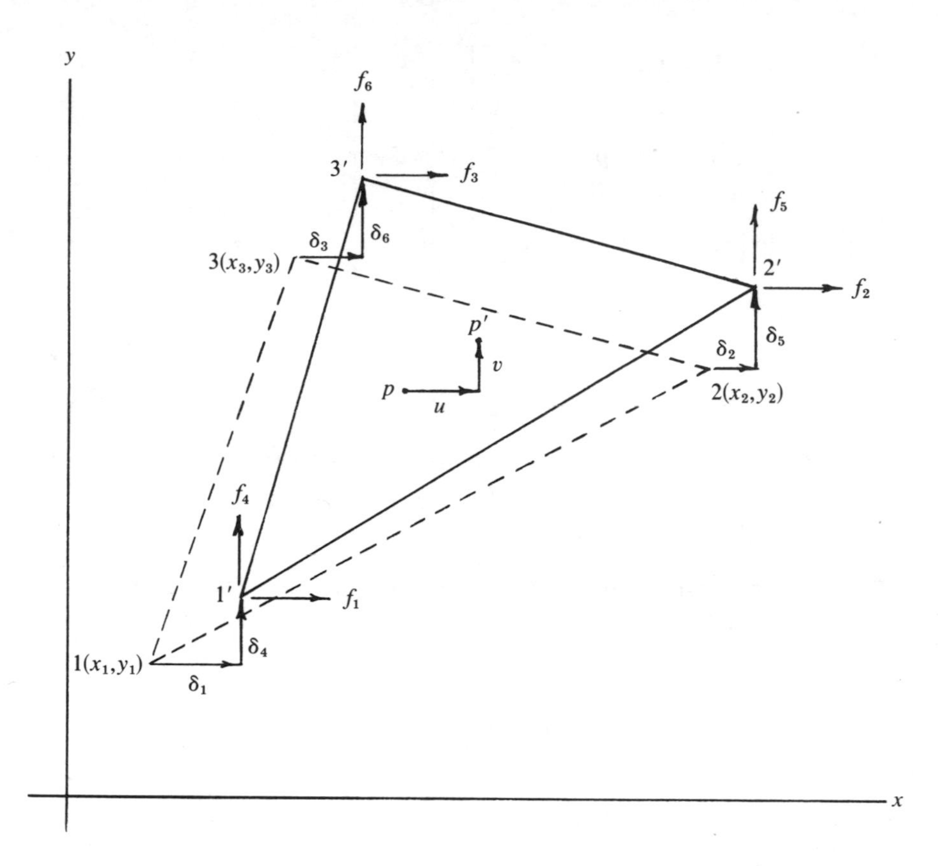

Figure 2-2. Displacement of Triangular Element

$$\begin{Bmatrix} u \\ v \end{Bmatrix} = \begin{bmatrix} 1 & x & y & 0 & 0 & 0 \\ 0 & 0 & 0 & 1 & x & y \end{bmatrix} \boldsymbol{\alpha} \qquad (2.2)$$

When $\boldsymbol{\alpha}$ is known, these expressions will give displacements at all locations in the plate. But it is known that

$$u_{(x=x_1)} = \delta_1 \quad \text{or} \quad [1 \;\; x_1 \;\; y_1 \;\; 0 \;\; 0 \;\; 0]\boldsymbol{\alpha} = \delta_1$$

The other five known displacements lead to similar expressions. These can be written as follows:

$$\begin{bmatrix} 1 & x_1 & y_1 & 0 & 0 & 0 \\ 1 & x_2 & y_2 & 0 & 0 & 0 \\ 1 & x_3 & y_3 & 0 & 0 & 0 \\ 0 & 0 & 0 & 1 & x_1 & y_1 \\ 0 & 0 & 0 & 1 & x_2 & y_2 \\ 0 & 0 & 0 & 1 & x_3 & y_3 \end{bmatrix} \begin{Bmatrix} \alpha_1 \\ \alpha_2 \\ \alpha_3 \\ \alpha_4 \\ \alpha_5 \\ \alpha_6 \end{Bmatrix} = \begin{Bmatrix} \delta_1 \\ \delta_2 \\ \delta_3 \\ \delta_4 \\ \delta_5 \\ \delta_6 \end{Bmatrix} \tag{2.3}$$

or,

$$\mathsf{A}\boldsymbol{\alpha} = \boldsymbol{\delta} \tag{2.4}$$

The solution for the unknown coefficients is given by

$$\boldsymbol{\alpha} = \mathsf{A}^{-1}\boldsymbol{\delta} \tag{2.5}$$

The inverse can be obtained by operating directly on A, but it is more economical to take advantage of the special properties of A and partition it into

$$\mathsf{A} = \begin{bmatrix} \bar{\mathsf{A}} & 0 \\ 0 & \bar{\mathsf{A}} \end{bmatrix}$$

and invert only $\bar{\mathsf{A}}$, which in this case is such a simple operation that it can be done by hand.

Then

$$\mathsf{A}^{-1} = \begin{bmatrix} \bar{\mathsf{A}}^{-1} & 0 \\ 0 & \bar{\mathsf{A}}^{-1} \end{bmatrix} \tag{2.6}$$

The coefficients in (2.2) can now be eliminated by substituting from (2.5) to get

$$\begin{Bmatrix} u \\ v \end{Bmatrix} = \begin{bmatrix} 1 & x & y & 0 & 0 & 0 \\ 0 & 0 & 0 & 1 & x & y \end{bmatrix} \mathsf{A}^{-1}\boldsymbol{\delta} \tag{2.7}$$

We will now pause to examine the implications of our choice of displacement functions. Referring to Figure 2-2, let us consider that all coordinates of nodes and all nodal displacements are known, then in (2.7) all coefficients in the polynomials are known. If we consider points on one side, say 1-2, the x and y in the polynomial are linearly related so that for points at varying distances from 1 toward 2 the u varies linearly as does also the v. Taking components of displacement normal and tangential to side 1-2, these also vary linearly, the constants in the linear function being determined by the displacements of nodes 1 and 2. On both sides of the interface 1-2, the displacements are determined by the same nodal displacements; hence adjacent points on opposite sides of the interface will have identical normal and tangential displacements. Therefore, at all points along the imaginary parting line 1-2 there is no relative motion that would either open a crack or cause relative sliding. This means that compatible displacements at the nodes, combined with the chosen displacement functions, ensure compatibility along all edges of the triangles and the gaps between adjacent triangles, as shown in Figure 2-1b, do not occur.

Strains in a Triangular Element

The displacement functions in the assumed form are not of much use to us. However they will lead us to a determination of strains and thence to our real objective, stress.

In the x,y plane, strain and displacement are related by

$$\varepsilon_{xx} = \frac{\partial u}{\partial x} \qquad \varepsilon_{yy} = \frac{\partial v}{\partial y} \qquad \varepsilon_{xy} = \frac{\partial u}{\partial y} + \frac{\partial v}{\partial x}$$

which can be verified by reference to a text on the theory of elasticity [12]. These expressions can be put into matrix form

$$\boldsymbol{\varepsilon} = \begin{Bmatrix} \varepsilon_{xx} \\ \varepsilon_{yy} \\ \varepsilon_{xy} \end{Bmatrix} = \begin{bmatrix} \frac{\partial}{\partial x} & 0 \\ 0 & \frac{\partial}{\partial y} \\ \frac{\partial}{\partial y} & \frac{\partial}{\partial x} \end{bmatrix} \begin{Bmatrix} u \\ v \end{Bmatrix}$$

When displacements are substituted from (2.7) this becomes

$$\boldsymbol{\varepsilon} = \begin{bmatrix} \frac{\partial}{\partial x} & 0 \\ 0 & \frac{\partial}{\partial y} \\ \frac{\partial}{\partial y} & \frac{\partial}{\partial x} \end{bmatrix} \begin{bmatrix} 1 & x & y & 0 & 0 & 0 \\ 0 & 0 & 0 & 1 & x & y \end{bmatrix} \mathsf{A}^{-1}\boldsymbol{\delta}$$

Since there are no variables in A^{-1} nor in $\boldsymbol{\delta}$, the differentiations have no effect on these quantities, and the differential operations lead to

$$\boldsymbol{\varepsilon} = \begin{bmatrix} 0 & 1 & 0 & 0 & 0 & 0 \\ 0 & 0 & 0 & 0 & 0 & 1 \\ 0 & 0 & 1 & 0 & 1 & 0 \end{bmatrix} \mathsf{A}^{-1}\boldsymbol{\delta}$$

Let

$$\mathsf{B} = \begin{bmatrix} 0 & 1 & 0 & 0 & 0 & 0 \\ 0 & 0 & 0 & 0 & 0 & 1 \\ 0 & 0 & 1 & 0 & 1 & 0 \end{bmatrix}$$

then

$$\boldsymbol{\varepsilon} = \mathsf{BA}^{-1}\boldsymbol{\delta} \tag{2.8}$$

Equation (2.8) gives the strains at any point $p(x, y)$ in the triangular element. But inspection of the terms in the equation shows that x and y do not appear and that all entries are constants. Consequently, the three strains have the same values at all points in the triangle. For this reason, the triangle is often referred to as the "constant strain triangle" although it might equally well be described as a "linearly varying displacement triangle."

Stress in a Triangular Element

In elementary mechanics of materials, the equations for strain in an isotropic material were established as

$$\varepsilon_{xx} = \frac{1}{E}\sigma_{xx} - \frac{\nu}{E}\sigma_{yy}$$

$$\varepsilon_{yy} = -\frac{\nu}{E}\sigma_{xx} + \frac{1}{E}\sigma_{yy}$$

$$\varepsilon_{xy} = \frac{2(1+\nu)}{E}\sigma_{xy}$$

which may be expressed in matrix form as

$$\begin{Bmatrix} \varepsilon_{xx} \\ \varepsilon_{yy} \\ \varepsilon_{xy} \end{Bmatrix} = \begin{bmatrix} \frac{1}{E} & -\frac{\nu}{E} & 0 \\ -\frac{\nu}{E} & \frac{1}{E} & 0 \\ 0 & 0 & \frac{2(1+\nu)}{E} \end{bmatrix} \begin{Bmatrix} \sigma_{xx} \\ \sigma_{yy} \\ \sigma_{xy} \end{Bmatrix}$$

The equations for stress can be obtained from these by inversion which gives

$$\begin{Bmatrix} \sigma_{xx} \\ \sigma_{yy} \\ \sigma_{xy} \end{Bmatrix} = \frac{E}{1-\nu^2} \begin{bmatrix} 1 & \nu & 0 \\ \nu & 1 & 0 \\ 0 & 0 & \frac{1-\nu}{2} \end{bmatrix} \begin{Bmatrix} \varepsilon_{xx} \\ \varepsilon_{yy} \\ \varepsilon_{xy} \end{Bmatrix}$$

Let

$$\mathsf{D} = \frac{E}{1-\nu^2} \begin{bmatrix} 1 & \nu & 0 \\ \nu & 1 & 0 \\ 0 & 0 & \frac{1-\nu}{2} \end{bmatrix} \qquad (2.9)$$

Then,

$$\boldsymbol{\sigma} = \mathsf{D}\boldsymbol{\varepsilon} \qquad (2.10)$$

Substituting from (2.8) gives

$$\boldsymbol{\sigma} = \mathsf{DBA}^{-1}\boldsymbol{\delta} \qquad (2.11)$$

This equation enables us to calculate the stresses in a triangular element provided we have the elastic properties, the coordinates of the corners, and the displacements of the corners. It is observed that there are no variables in the right-hand side of (2.11) and hence the stresses are constant throughout the element.

Stiffness of Triangular Element

We will now follow the procedure that was used in Chapter 1 to establish the stiffness of the element. Imagine that the element in Figure 2-2 has been given the displacements $\boldsymbol{\delta}$ and we would like to be able to determine the forces $\mathbf{f}$ necessary to hold the element in the strained state. This could be done if we knew K, the stiffness matrix in

$$\mathbf{f} = \mathsf{K}\boldsymbol{\delta} \qquad (2.1)$$

In the state described above, the existing stresses are given by (2.11). Let us superimpose small virtual displacements $\boldsymbol{\delta}^*$ on the existing configuration. The virtual displacements must be small enough so that the forces do not change significantly during the displacement. The work done by the forces is then given by

$$w_E = \delta_1^* f_1 + \delta_2^* f_2 + \delta_3^* f_3 + \delta_4^* f_4 + \delta_5^* f_5 + \delta_6^* f_6 = \boldsymbol{\delta}^{*T}\mathbf{f} \qquad (2.12)$$

During the virtual displacement the strain is given by (2.8) as

$$\boldsymbol{\varepsilon}^* = \mathsf{B}\,\mathsf{A}^{-1}\boldsymbol{\delta}^* \qquad (2.13)$$

The stress, given by (2.11), which is substantially constant, does work during the virtual strain of an amount that can be calculated for an incremental volume as

$$dw_I = \varepsilon_{xx}^* \sigma_{xx}\, dv + \varepsilon_{yy}^* \sigma_{yy}\, dv + \varepsilon_{xy}^* \sigma_{xy}\, dv = \boldsymbol{\varepsilon}^{*T}\boldsymbol{\sigma}\, dv$$

Substituting from (2.11) and (2.13) the above expression becomes

$$dw_I = [\mathsf{BA}^{-1}\boldsymbol{\delta}^*]^T\mathsf{DBA}^{-1}\boldsymbol{\delta}\, dv = \boldsymbol{\delta}^{*T}[\mathsf{A}^{-1}]^T\mathsf{B}^T\mathsf{DBA}^{-1}\boldsymbol{\delta}\, dv$$

In the whole element, the internal work done is

$$w_I = \int_{\mathrm{vol}} \boldsymbol{\delta}^{*T}[\mathsf{A}^{-1}]^T\mathsf{B}^T\mathsf{DBA}^{-1}\boldsymbol{\delta}\, dv$$

Since all quantities in the matrices are constant, the integration needs to be done only on dv which gives the volume of the element, hence

$$w_I = \boldsymbol{\delta}^{*T}[\mathsf{A}^{-1}]^T\mathsf{B}^T\mathsf{DBA}^{-1}\,\mathrm{vol}\,\boldsymbol{\delta}$$

Since there is no loss of energy in this process, the external energy can be equated to the internal energy giving

$$\boldsymbol{\delta}^{*T}\mathbf{f} = \boldsymbol{\delta}^{*T}[\mathsf{A}^{-1}]^T\mathsf{B}^T\mathsf{DBA}^{-1}\,\mathrm{vol}\,\boldsymbol{\delta}$$

which leads to

$$\mathbf{f} = [\mathsf{A}^{-1}]^T\mathsf{B}^T\mathsf{DBA}^{-1}\,\mathrm{vol}\,\boldsymbol{\delta}$$

This is of the same form as (2.1) with

$$\mathsf{K} = [\mathsf{A}^{-1}]^T\mathsf{B}^T\mathsf{DBA}^{-1}\,\mathrm{vol} \qquad (2.14)$$

Every term on the right side of (2.14) is known, so we can calculate the stiffness matrix. For the particular case under consideration, we could work a general solution to all triangular element stiffnesses. This has been done and the solution is to be found in many places and will not be repeated here (see, for example, Clough [2] or Turner, Clough, Martin, and Topp [14]). In practice it is better to generate the stiffness of each element from (2.14) as it is needed since the programming required to fill in K directly is quite tedious.

Stresses in an In-Plane Loaded Plate

As the stiffness of each triangular element, K, is determined it can be added to the general stiffness matrix K_A in a manner similar to that of the bar elements of Chapter 1. When all triangles in a plate have been considered, the K_A matrix will represent the total stiffness of the plate in

$$\mathbf{F} = \mathsf{K}_A\mathbf{U}$$

Hereafter the subscript, A, will be omitted for simplicity, the context being sufficient to indicate which stiffness is being referred to. Thus the general stiffness equation becomes

$$\mathbf{F} = \mathsf{K}\mathbf{U} \tag{2.15}$$

For a loaded plate with a given supporting system, part of the **F** and part of the **U** components are known. This is the same situation that we had in the pin-jointed structure and the methods for finding the unknown components are identical. With **U** determined, the components can be selected for substitution into (2.11) and thus the stress in any element obtained.

Nodes Constrained to Move Along Inclined Lines

In the theory that has been developed, the displacement components were considered to be parallel to the x and y coordinate axes; hence by specifying one displacement component at a node and leaving the other component as an unknown, the node can be constrained to move in the horizontal or vertical direction. This enables the user to take advantage of axes of symmetry and thereby reduce the size of the problem to be solved. For example, the plate shown in Figure 2-3a could be solved by dealing with only that part shown in Figure 2-3b. If the problem had been to solve the plate shown in Figure 2-4a, one quadrant could still be used to obtain the solution, but this would fail to take advantage of the symmetry that exists about the diagonals. If it is possible to constrain the nodes to move only along inclined lines, the problem can be reduced to that of Figure 2-4b. This would save both computer time and the space required to solve the problem or, for equal cost, would allow the engineer to subdivide the plate into smaller triangles thus improving accuracy.

In practice, it is usually required to treat several nodes that are guided to move along inclined lines. The treatment that follows deals with only one node, but for several guided nodes the process is merely repeated. Usually the displacement normal to the guiding line is zero, but to make the

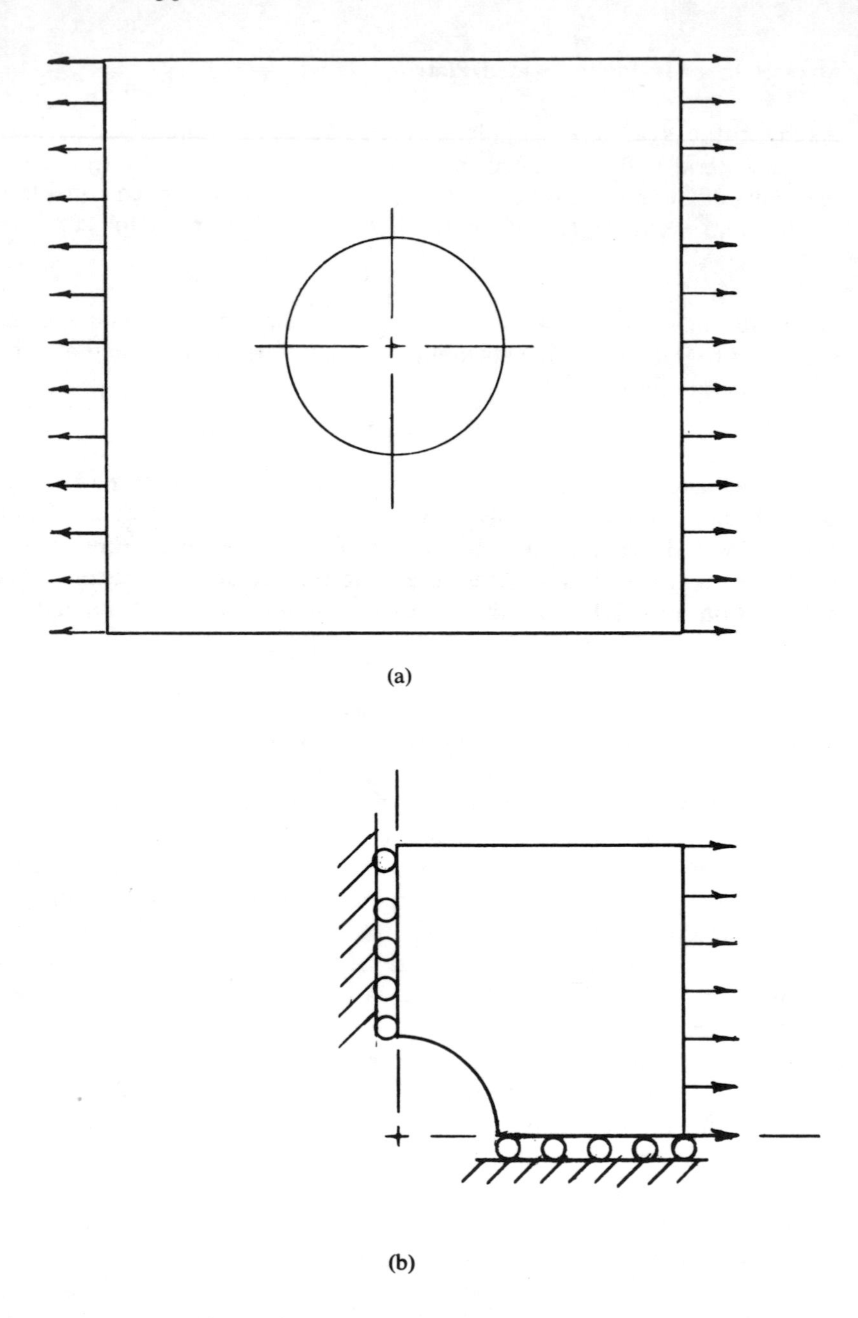

Figure 2-3. Problem Reduction Using Axes of Symmetry

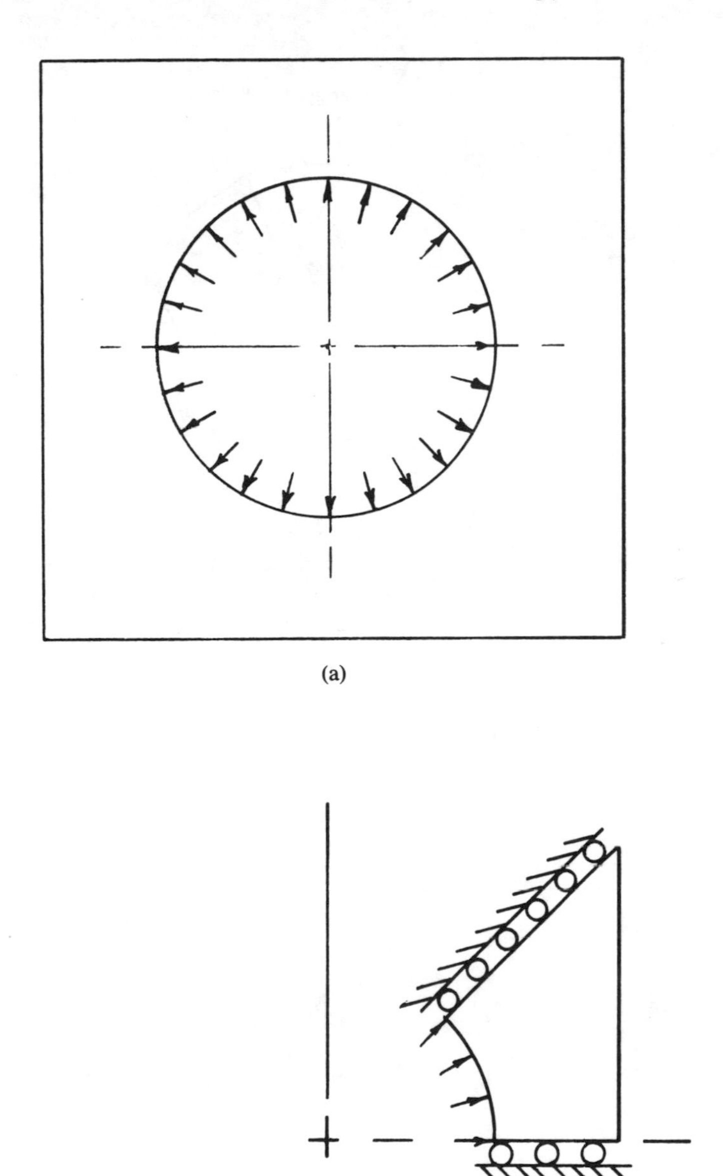

Figure 2-4. Problem Reduction Using Axes of Symmetry

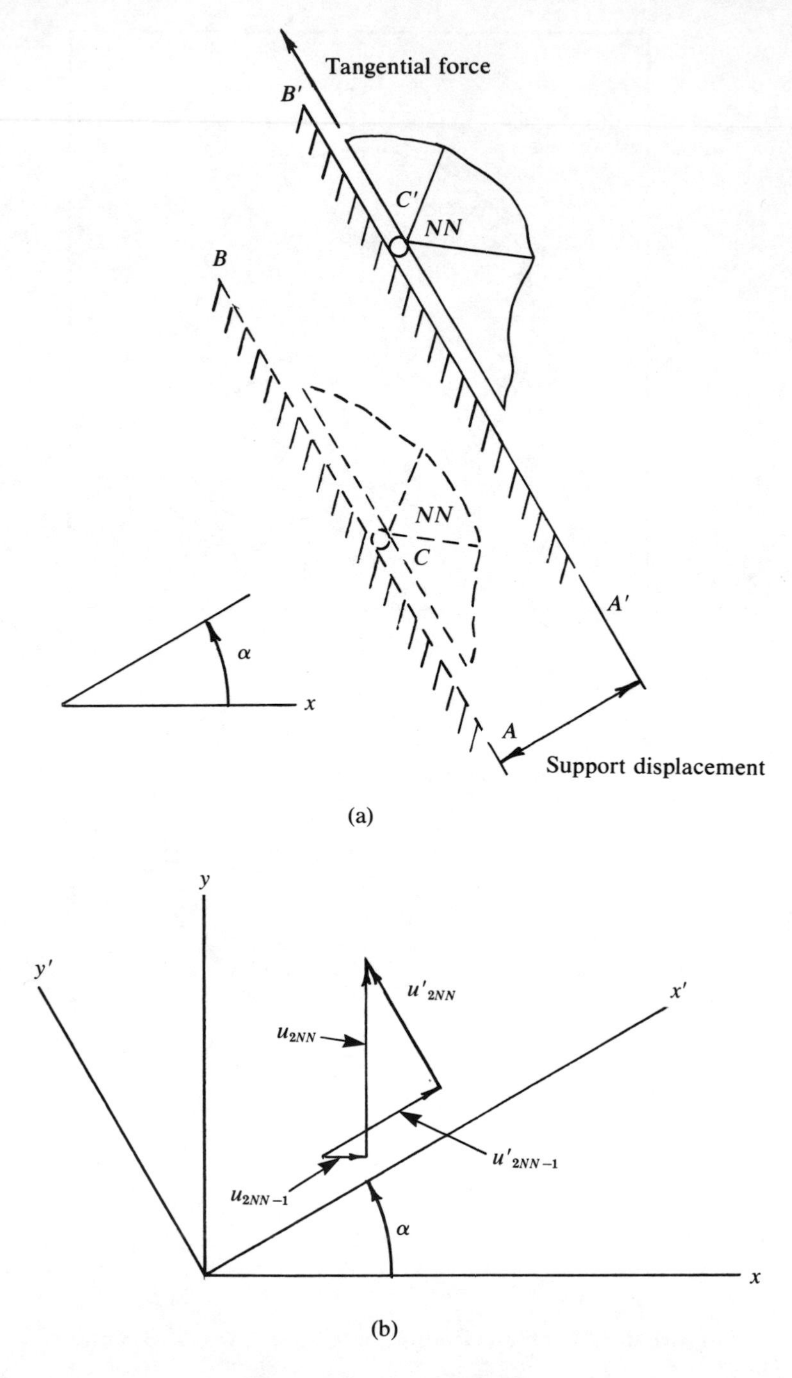

Figure 2-5. Guided Node

following more general, a known normal displacement will be considered and a known tangential force provided for.

Consider the node, number NN, in Figure 2-5a. The unloaded position of the node is given by C. After loading it moves to C', the support having moved a known amount in the direction normal to AB. To deal with this displacement state and the attendant forces so that known and unknown components may be kept separated, it will be necessary to consider the axes at node C to be locally rotated through an angle, α, as shown in Figure 2-5b. The relationship between the two displacement systems can be seen from Figure 2-5b to be

$$\begin{Bmatrix} u_{2NN-1} \\ u_{2NN} \end{Bmatrix} = \begin{bmatrix} \cos\alpha & -\sin\alpha \\ \sin\alpha & \cos\alpha \end{bmatrix} \begin{Bmatrix} u'_{2NN-1} \\ u'_{2NN} \end{Bmatrix}$$

or,

$$\mathbf{u} = \mathsf{T}\mathbf{u}' \tag{2.16}$$

where

$$\mathsf{T} = \begin{bmatrix} \cos\alpha & -\sin\alpha \\ \sin\alpha & \cos\alpha \end{bmatrix} \tag{2.17}$$

Hence,

$$\mathbf{u}' = \mathsf{T}^{-1}\mathbf{u} = \mathsf{T}^{T}\mathbf{u} \tag{2.18}$$

since $\mathsf{T}^{-1} = \mathsf{T}^{T}$ in this particular case.

Similarly for the two components of force at the guided node:

$$\mathbf{f} = \mathsf{T}\mathbf{f}' \tag{2.19}$$

and

$$\mathbf{f}' = \mathsf{T}^{T}\mathbf{f} \tag{2.20}$$

However, because these two pairs of components are part of larger displacement and force vectors in

$$\mathbf{F} = \mathsf{K}\mathbf{U} \tag{2.15}$$

they can not be treated alone.

When the axes are rotated locally at node C, only the components $2NN$ and $2NN-1$ in $\mathbf{F}$ and $\mathbf{U}$ are altered. From (2.20) we see that the new force vector is given by,

$$\mathbf{F}' = \begin{bmatrix} \mathsf{I}_1 & 0 & 0 \\ 0 & \mathsf{T}^T & 0 \\ 0 & 0 & \mathsf{I}_2 \end{bmatrix} \mathbf{F} \tag{2.21}$$

Where I_1 is a $(2NN - 2)$ by $(2NN - 2)$ unit matrix, T is a 2 by 2 matrix, and I_2 is large enough to make the array NDF by NDF.

Similarly, using (2.16)

$$\mathbf{U} = \begin{bmatrix} \mathsf{I}_1 & 0 & 0 \\ 0 & \mathsf{T} & 0 \\ 0 & 0 & \mathsf{I}_2 \end{bmatrix} \mathbf{U}' \tag{2.22}$$

Premultiplying both sides of (2.15) by

$$\begin{bmatrix} \mathsf{I}_1 & 0 & 0 \\ 0 & \mathsf{T}^T & 0 \\ 0 & 0 & \mathsf{I}_2 \end{bmatrix}$$

gives

$$\begin{bmatrix} \mathsf{I}_1 & 0 & 0 \\ 0 & \mathsf{T}^T & 0 \\ 0 & 0 & \mathsf{I}_2 \end{bmatrix} \mathbf{F} = \begin{bmatrix} \mathsf{I}_1 & 0 & 0 \\ 0 & \mathsf{T}^T & 0 \\ 0 & 0 & \mathsf{I}_2 \end{bmatrix} \mathsf{K}\mathbf{U} \tag{2.23}$$

Substituting (2.21) and (2.22) into (2.23) gives

$$\mathbf{F}' = \begin{bmatrix} \mathsf{I}_1 & 0 & 0 \\ 0 & \mathsf{T}^T & 0 \\ 0 & 0 & \mathsf{I}_2 \end{bmatrix} \mathsf{K} \begin{bmatrix} \mathsf{I}_1 & 0 & 0 \\ 0 & \mathsf{T} & 0 \\ 0 & 0 & \mathsf{I}_2 \end{bmatrix} \mathbf{U}'$$

or

$$\mathbf{F}' = \mathsf{K}'\mathbf{U}' \tag{2.24}$$

Where,

$$\mathsf{K}' = \begin{bmatrix} \mathsf{I}_1 & 0 & 0 \\ 0 & \mathsf{T}^T & 0 \\ 0 & 0 & \mathsf{I}_2 \end{bmatrix} \mathsf{K} \begin{bmatrix} \mathsf{I}_1 & 0 & 0 \\ 0 & \mathsf{T} & 0 \\ 0 & 0 & \mathsf{I}_2 \end{bmatrix} \tag{2.25}$$

Hence we can deal with axes at one node that are locally rotated, provided that the stiffness matrix is modified in accordance with (2.25).

The operations in (2.25) can be greatly simplified if advantage is taken of the special nature of the unit and the null matrix. Consider the K matrix partitioned to conform to I_1 which is $(2NN - 2)$ by $(2NN - 2)$ and T which is 2×2 as follows:

$$\mathsf{K} = \begin{array}{c} 2NN-2 \\ 2 \\ \\ \end{array} \left[\begin{array}{c|c|c} K_{11} & K_{12} & K_{13} \\ \hline K_{21} & K_{22} & K_{23} \\ \hline K_{31} & K_{32} & K_{33} \end{array} \right]$$

(column widths: $2NN - 2$, 2)

Performing the operations of (2.25) gives

$$\mathsf{K}' = \begin{bmatrix} K_{11} & K_{12}T & K_{13} \\ T^T K_{21} & T^T K_{22} T & T^T K_{23} \\ K_{31} & K_{32}T & K_{33} \end{bmatrix} \tag{2.26}$$

(middle row and middle column width: 2)

which shows that only two rows and two columns are altered in the transformation resulting from local rotation of axes at a guided node.

When several nodes are so constrained, the operation given in (2.26) is repeated for each constrained node. The altered matrix K', the known components of $\mathbf{U}'$ and $\mathbf{F}'$ are then used to solve for all $\mathbf{U}'$. Since stress can readily be calculated from x and y components of displacement but not from a mixed set of components, the $\mathbf{U}'$ components are used to determine all $\mathbf{U}$ by repeated application of equation (2.22). Then by recalling the original K, and using $\mathbf{U}$, the original $\mathbf{F}$ is calculated.

Stiffness Matrix for Any Finite Element

The triangular element has been treated so that its stiffness matrix can be calculated and stresses in each element of an assembly of triangles can be found. We will now solve the same problem but without reference to any particular element, thus arriving at a general solution. While going through this process we will add another condition, that is, we will make provision for strain that is not the result of stress and will thus make it possible to treat cases involving thermal strain, creep, etc.

Consider that the element, irrespective of its shape, degrees of freedom, or number of components in its stress and strain vectors, has a strain $\boldsymbol{\varepsilon}_0$ which is not related to load, that is, the strains $\boldsymbol{\varepsilon}_0$ would exist if there were no constraints on the element and it could freely change size from temperature change or for other reasons. This strain $\boldsymbol{\varepsilon}_0$ will cause certain displacements which we will not ignore. When force $\mathbf{f}$ is applied additional displacements will occur giving a total displacement represented by displacement functions in $\mathbf{u}$. Total strains can be found from the displacement function by performing certain differential operations given in an operator matrix, $\boldsymbol{\Delta}$. That is,

$$\boldsymbol{\varepsilon} = \boldsymbol{\Delta}\mathbf{u} \tag{2.27}$$

But the elastic strain is given by $\boldsymbol{\varepsilon} - \boldsymbol{\varepsilon}_0$, and hence the stress is

$$\boldsymbol{\sigma} = \mathsf{D}\{\boldsymbol{\varepsilon} - \boldsymbol{\varepsilon}_0\} \tag{2.28}$$

The displacement functions $\mathbf{u}$ are polynomials and can be written

$$\mathbf{u} = \mathsf{P}\boldsymbol{\alpha} \tag{2.29}$$

where P contains polynomial terms and $\boldsymbol{\alpha}$ contains the constants in the polynomial function. Since the displacement function must give the displacements at the nodes, $\boldsymbol{\delta}$, when the coordinates of the nodes are substituted into P, (2.29) gives

$$\boldsymbol{\delta} = \mathsf{A}\boldsymbol{\alpha} \tag{2.30}$$

where A comes from P evaluated at the nodes. Solving (2.30) gives the coefficients in the polynomial $\boldsymbol{\alpha} = \mathsf{A}^{-1}\boldsymbol{\delta}$. Substituting this into (2.29) and then into (2.27) gives

$$\boldsymbol{\varepsilon} = \boldsymbol{\Delta}\mathsf{P}\mathsf{A}^{-1}\boldsymbol{\delta} \tag{2.31}$$

The operations of $\boldsymbol{\Delta}$ can be performed on P giving a new matrix B:

$$B = \Delta P \tag{2.32}$$

Substituting into (2.31),

$$\varepsilon = BA^{-1}\delta \tag{2.33}$$

which, when substituted into (2.28) gives

$$\sigma = DBA^{-1}\delta - D\varepsilon_0 \tag{2.34}$$

At this stage we have an element that has its nodes displaced by δ and is held in this position by some, as yet unknown, force **f**. From this state let us give the element a virtual displacement δ^* which will cause additional strain ε^* within the element. From (2.33) we then have

$$\varepsilon^* = BA^{-1}\delta^* \tag{2.35}$$

The additional strain energy stored by the element during this step in an incremental volume is $\varepsilon^{*T}\sigma\, dv$, which gives for the whole element by substituting from (2.34) and (2.35):

Increase in strain energy

$$= \int_{vol} [BA^{-1}\delta^*]^T [DBA^{-1}\delta - D\varepsilon_0]\, dv$$

$$= \int_{vol} [\delta^{*T} [A^{-1}]^T B^T][DBA^{-1}\delta - D\varepsilon_0]\, dv \tag{2.36}$$

When this virtual displacement is imposed, the external forces do an amount of work given by

$$\text{External work} = \delta^{*T}\mathbf{f}$$

Equating external work to the increase in strain energy and removing constant terms from the integration process gives

$$\delta^{*T}\mathbf{f} = \delta^{*T} [A^{-1}]^T \int B^T[DBA^{-1}\delta - D\varepsilon_0]\, dv$$

This can be arranged in the form

$$\mathbf{f} = [A^{-1}]^T \int B^T DB\, dv\, A^{-1}\delta - [A^{-1}]^T \int B^T D\varepsilon_0\, dv \tag{2.37}$$

And if we let

$$K = [A^{-1}]^T \int B^T DB\, dv\, A^{-1} \tag{2.38}$$

then,

$$\mathbf{f} = K\delta - [A^{-1}]^T \int B^T D\varepsilon_0\, dv \tag{2.39}$$

Equation (2.39) relates force to displacement through K which is the stiffness matrix determined by (2.38). If we let

$$\mathbf{f}_0 = [\mathsf{A}^{-1}]^T \int \mathsf{B}^T \mathsf{D} \boldsymbol{\varepsilon}_0 \, dv \tag{2.40}$$

then equation (2.39) can be written

$$\mathbf{f} + \mathbf{f}_0 = \mathsf{K}\boldsymbol{\delta} \tag{2.41}$$

In this expression, the components of **f** are the actual loads that are imposed on the element and $\mathbf{f}_0$ components are ficticious forces being in fact the nodel forces that would produce the strain $\boldsymbol{\varepsilon}_0$, but this does not need to be recognized. To solve a problem we merely calculate $\mathbf{f}_0$ for each element, treat it as though it is a real load, and add its components to the proper components of the force vector **F**. This can be done conveniently while the stiffness matrix is being generated.

After all displacements for the assembly have been determined, those that apply to each element are selected in turn and the stress found by substitution into (2.34).

A Program to Determine Stresses in a Plate Due to Temperature and In-Plane Loads

The methods described in this chapter have been programmed in SP23B to solve a wide range of plane stress problems. While reading this section, reference should be repeatedly made to Figure 2-6, SP23B Flow Chart; Figure 2-7, SP23B Fortran Statements; Figure 2-8, Instructions for SP23B Data Deck Preparation; and the computer output as given in the examples of the next section. Many of the program steps are quite similar to those already described in detail for PT10B in Chapter 1. These will be repeated here for completeness but less detail will be given.

When SP23B execution starts, the computer prints a main title and a subtitle which is read from the first data card.

The next data card contains: Young's Modulus; Poisson's ratio; initial temperature; and the coefficient of thermal expansion. All of these values are taken as applying to all parts of the body. The initial temperature is that at which the unloaded body is completely unstressed.

The node location data is specified by punching one card for each node containing the node number and the coordinates of the nodes.

Element data is presented by punching one card for each element containing the element number, the three node numbers at the corners of the triangle, the element thickness, and the final temperature. The thickness as well as the temperature may change from element to element. Since many problems are solved with unit thickness the program is written so that

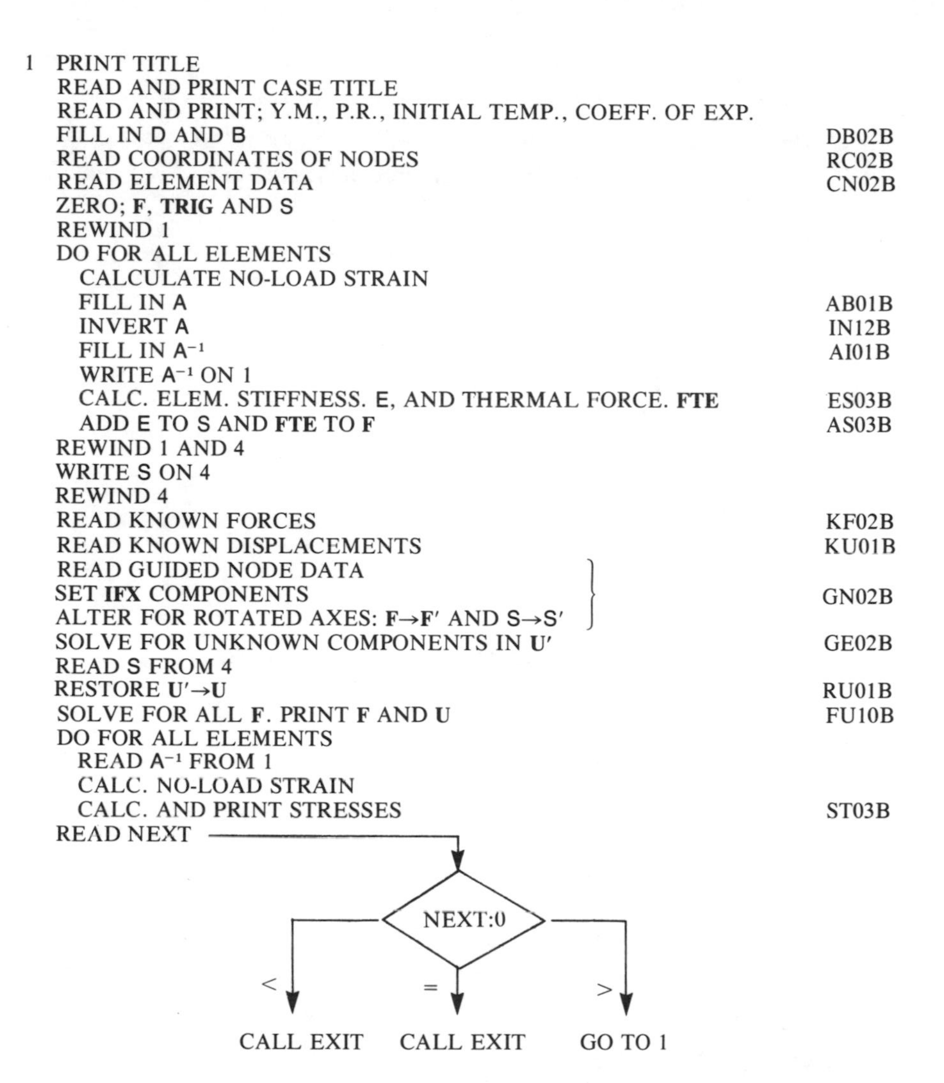

Figure 2-6. SP23B Flow Chart

1.0 is entered when the thickness is read as zero. This effects a saving in punching in a great many cases. The end of the element data cards is indicated by a blank card.

Having sufficient data, the program proceeds to calculate for each element, in turn, its stiffness and the force that is equivalent to the thermal expansion. These are accumulated in **S** and **F** so that when all elements have been treated the **S** matrix is complete and **F** contains the total thermal equivalent load. As each element is dealt with the matrix A^{-1} is determined

```
      *****      MAIN SP23B       *****            AUG 20 1971
      STRESSES IN A PLATE DUE TO TEMPERATURE AND IN-PLANE LOADS
   BY FINITE ELEMENT METHODS  USING CONSTANT STRAIN
   TRIANGLES                                W. H. BOWES
      DIMENSION E(6,6),B(3,6),D(3,3),AB(3,3),AI(6,6),FTE(6),NAME(20)
      DIMENSION S(30,120)
      DIMENSION F(120),U(120),IFX(120),TRIG(120),XY(2,60)
      DIMENSION NCON(3,100),T(100),DF(100)
      LBAND=30
    1 WRITE(3,111)
      WRITE(3,112)
      READ(2,100)NAME
      WRITE(3,101)NAME
      READ(2,110)YM,GNU,DFI,CE
      WRITE(3,104)YM,GNU
      WRITE(3,105)DFI,CE
      CALL DB02B(YM,GNU,D,B)
      CALL RC02B(2,XY,NDF)
      CALL CN02B(LBAND,NCON,T,DF,NE,NBAND)
      DO 5 J=1,NDF
      F(J)=0.
      TRIG(J)=0.
      DO 5 I=1,NBAND
    5 S(I,J)=0.
      REWIND 1
      DO 25 IE=1,NE
      ET=CE*(DF(IE)-DFI)
      CALL AB01B(IE,NCON,XY,T,AB,VOL)
      CALL IN12B(AB,3)
      CALL AI01B(AB,AI)
      WRITE(1) AI
      CALL ES03B(AI,B,D,VOL,ET,E,FTE)
   25 CALL AS03B(IE,NCON,E,FTE,S,F,2,3,6,LBAND,1)
      END FILE 1
      REWIND 1
      REWIND 4
      DO 26 I=1,NBAND
   26 WRITE(4)(S(I,J),J=1,NDF)
      END FILE 4
      REWIND 4
      CALL KF02B(F)
      CALL KU01B(U,IFX,NDF)
      CALL GN02B(S,J,F,IFX,NBAND,LBAND,TRIG)
      CALL GE02B(F,J,S,IFX,NBAND,NDF,LBAND)
      DO 30 I=1,NBAND
   30 READ(4) (S(I,J),J=1,NDF)
      CALL RU01B(U,TRIG,NDF)
      CALL FU10B(S,J,NBAND,NDF,2,LBAND)
      WRITE(3,106)
      DO 45 IE=1,NE
      READ(1) AI
      ET=CE*(DF(IE)-DFI)
   45 CALL ST03B(D,B,AI,ET,U,NCON,IE)
      READ(2,108)NEXT
      IF(NEXT)47,47,1
   47 CALL EXIT
  100 FORMAT(20A4)
  101 FORMAT('0CASE TITLE ---  ',20A4)
  104 FORMAT('0YOUNGS MODULUS =',E10.3,'  POISSONS RATIO =',F6.3)
  105 FORMAT(' INITIAL TEMP =',F6.1,'   COEF.OF EXP.=',E10.3)
  106 FORMAT('0ELEM.NO.     SXX           SYY           SXY          THETA      PS1
     1       PS2')
  108 FORMAT(I5)
  110 FORMAT(4F10.5)
  111 FORMAT('1                MAIN SP23B   AUG  20 1971')
  112 FORMAT(' STRESSES IN A PLATE DUE TO TEMPERATURE AND IN-PLANE LOADS
     1',/,' BY FINITE ELEMENT METHODS USING CONSTANT STRAIN TRIANGLES ')
      END
```

Figure 2-7. SP23B Fortran Statements

Data Deck:

A card containing the Case Title.

A card containing physical properties;

Young's modulus, Poisson's ratio, initial temperature, Coefficient of linear thermal expansion. Format: 4F10.5.

One card for each node containing;

Node number, x coordinate, y coordinate. Format: I5, 2F10.5

A blank card to indicate end of node data.

One card for each element containing;

Element number, Node numbers as apices, Thickness of element (= 1 by default), Final temperature. Format: 4I5, 2F10.5

A blank card to indicate end of element data.

One card for each known, nonzero load component containing;

Component number (Twice node number minus one for x component or twice node number for y component), Magnitude of force component.
Format: I5, F10.5
Note: It is not necessary to read in zero components.

A blank card to indicate end of load data.

One card for each known displacement component containing;

Component number, Displacement. Format: I5, F10.5.
Note: Zero displacements must be entered.

A blank card to indicate end of displacement data.

One card for each guided node containing;

Node number, Angle (alpha) in degrees defining direction of guiding plane, Normal displacement of guiding plane, Load tangential to guiding plane. Format: I5, 3F10.5
Note: Consider a set of axes, x′ and y′, rotated so that x′ is normal to the guiding plane. The angle of the x′ axis when measured counter-clockwise from the x axis gives alpha. Normal displacement is positive if the plane moves in the $+x'$ direction. The load is positive if it is in the $+y'$ direction.

A blank card to indicate end of guided node data.

A card containing NEXT. Format I5

To process another case, enter any positive integer and follow by case data deck prepared in accordance with all instructions given above. Use a blank card to CALL EXIT.

Figure 2-8. Instructions for SP23B Data Deck Preparation

and used. However, it is required later in the program when stresses are found by (2.34) and so is saved for that phase by writing it on tape or disc. Since it will also be necessary to recall S, it also is written into storage.

External loads are now accepted by reading load cards containing the component number and the magnitude of the force component. These are read until a blank card is encountered which causes the computer to begin reading known displacements. Component numbers and displacement components are read until another blank card is read. All of these known displacement components must be parallel to the coordinate axes. The vector **IFX** is used to record which components are known.

In the next phase, guided nodes are dealt with. For each node having a support that can be represented by a roller on a plane that is not parallel to either the x or y axis a guided node data card is punched. A card contains the node number, the inclination of the guiding plane, the displacement of the plane normal to its surface, and any load component that is parallel to the surface. As each card is read, **IFX** is altered to indicate that a certain component is known. The values of the trigonometric functions in (2.17) are determined and are saved in **TRIG** for further use. They are also used to alter **F** by (2.21) and also K (in reality S) by (2.25). A blank card indicates the end of the guided node data.

At this stage K and some components of **U** and some of **F** are known. As described in Chapter 1, the method of Payne and Irons is used to solve for the unknown displacements. However, some of these components, where there are guided nodes, refer to locally rotated axes. These are used with data in **TRIG** to determine the x and y components by (2.22). Using the restored **U** and the original S which is recalled from storage, all **F** components are calculated.

The program prints all **F** and **U** for the information of the user; but where there are thermal loads, the **F** values are difficult to interpret.

The program then goes on to calculate stresses in all elements making use of all the A^{-1} matrices that were stored earlier in the program. As well as the coordinate stresses, the program calculates and prints out the principle stresses and their directions.

At this stage the reader should get some experience in solving problems by the Finite Element Method and SP23B is recommended for this purpose. Some very simple problems having known solutions should be attempted first. If this is always done with all newly acquired programs a great reduction in wasted computer time will be made as almost invariably errors will be made in interpreting the instructions for data deck preparation. The introductory problems should have known solutions so that they can also serve as a check on the correctness of the program. When confidence has been established, the real problems can be attempted.

When SP23B is used to solve a problem in which the stress is constant, the solution will give displacements and stresses that are exact. If the stress varies it will be found that the displacements are quite accurate but the stresses may be very much in error. This is because the method is approximate; the approximation being introduced when we assumed that displacement within the element varied linearly, which implies that stress is constant throughout each element. If a case is imagined where the stress varies sharply, for example, near a hole in a plate, the element stresses which vary stepwise from element to element give a poor representation of the real stresses. This is especially true if the elements span regions in which there are large variations in stress. The obvious way to improve the results is by making the elements small in regions where the stress is likely to vary sharply. Of course, this must be done by intuition when the solution is not known in advance. It is in the realm of element size grading that the skill of the user becomes important. The user must decide on the total number of elements and their sizing. The grading decision can be avoided by using a large number of uniformly small elements but this leads to large matrices, perhaps beyond the capacity of the program and computer, and long execution time. Most users can not afford to sidestep the decision by this method. Even with skillful size grading, it is not uncommon to find that a thousand elements are required to attain the needed accuracy.

When the stresses have been calculated, there remains the question of interpretation. For any element, the stress obtained applies to all points in the element so at the interface between elements there is a step in stress level. Some very good results have been obtained by averaging the stresses of all elements that meet at a node and treating this as the stress at the nodal location. But there are many ways to weight the stresses in the averaging process and no single system seems to give consistently good results. It appears that the simplest interpretation scheme is the best, that is to consider that the stress in the element applies to the location of its centroid. This has the disadvantage of never giving stress at a boundary but boundary stresses can be approximated by some form of extrapolation.

Examples

The examples of a thin disc, first under the action of an internal pressure and secondly with a temperature variation, have been selected for this chapter to show the form of the computer output; to show the versatility of the method; and to give an indication of the accuracy obtainable. In order that an indication of the accuracy can be had, the computer analysis has been done with a very few elements and then repeated with a larger number of elements. The results are then compared, graphically, with the theoretical analysis.

Problem 2.1

For the thin disc shown in Figure 2-9 determine the radial and circumferential stresses by the Finite Element Method and compare the results with the theoretical solution.

Solution. In order to solve this problem by the Finite Element Method, consideration must first be given to subdivision of the problem into ele-

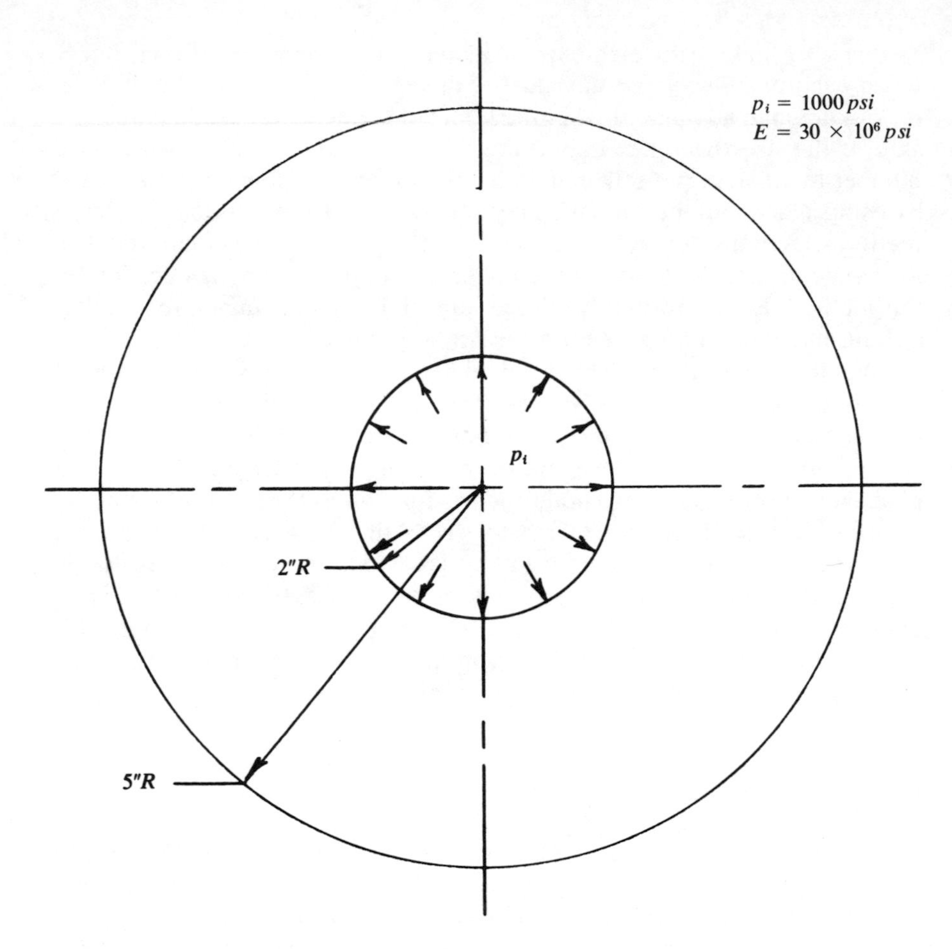

Figure 2-9. Thin Disc with Internal Pressure

ments. One could, of course, subdivide the full disc into elements but this would result in a large number of elements and a subsequent wastage in computer time and space. It is therefore worthwhile to recognize the polar symmetry of the problem and hence reduce the problem to one of finding the stresses in a wedge with the appropriate boundary conditions. Because there are an infinite number of axes of symmetry in this problem, almost any size of wedge could be chosen, a practical size—not too big or too small—has an included angle of say 15°. The next step is to choose the number of elements. In this example, we will first solve the problem with 4 elements and then with 12 elements in order to obtain some appreciation of the resulting increase in accuracy. The subdivision into elements for both cases is shown in Figure 2-10.

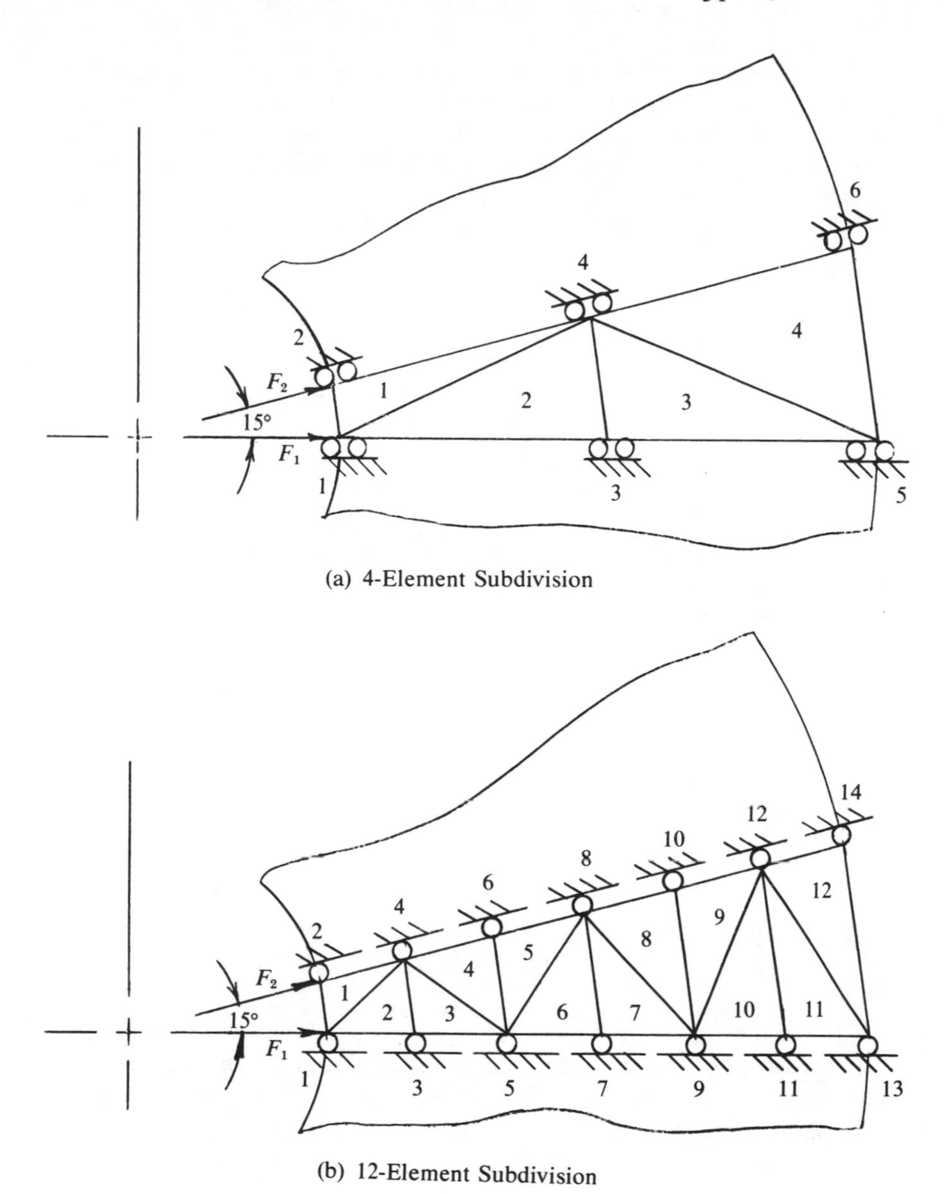

Figure 2-10. Element Subdivision for Thin Disc (a) 4-Element Subdivision (b) 12-Element Subdivision

The internal pressure loading is represented for the Finite Element Method by equivalent nodal point loads. In this case, since we are using a constant strain triangle, nodal point loads, statically equivalent to the internal pressure, will be applied as shown.

The boundary conditions are determined from the nature of the problem and because of symmetry, radial motion along any line of symmetry is permitted. The boundary constraint will therefore be: no normal displacement permitted along the sides of the wedge. This is depicted in Figure 2-10 by rollers placed at the nodes.

Solutions for the two cases are shown in Figures 2-11 and 2-12.

Problem 2.2

For the disc, with temperature varying from 300°F on the inside to 200°F on the outside, shown in Figure 2-14 find the radial and circumferential stresses using the Finite Element Method. Vary the number of elements and compare the results with the theoretical solution.

Solution. The polar symmetrical nature of the problem as with Problem 2.1 permits the problem to be solved by use of a wedge rather than a full disc. Therefore the same subdivision of elements will be used in this problem as was used in Problem 2.1. Again, because of the polar symmetry, the boundary conditions are the same as in Problem 2.1.

The internal pressure, in this example, is zero and hence the radial nodal loads designated in Figure 2-10 as F_1 and F_2 will be zero. The temperature varies throughout the disc from 300°F at the inside to 200°F at the outside. This variation follows the relation

$$T = T_i + \frac{T_o - T_i}{\ln r_o/r_i} \ln \frac{r}{r_i}$$

which for this particular problem reduces to

$$T = -109.14 \ln r + 375.65$$

Because this program uses constant strain triangles, the temperature is considered to be constant throughout each element. The value of the temperature specified for each element is the value found at the centroid of the element using the above equation.

Results

The computer output for Problems 2-1 and 2-2 is shown in Figures 2-11 and 2-12 and Figures 2-15 and 2-16 respectively. It is interpreted graphically in Figures 2-13 and 2-17.

In Problem 2-1 with four elements, we observe in Figure 2-13 that the results generally agree with the theoretical answer showing a zig-zag pat-

```
          MAIN SP23B   AUG  20 1971
STRESSES IN A PLATE DUE TO TEMPERATURE AND IN-PLANE LOADS
BY FINITE ELEMENT METHODS USING CONSTANT STRAIN TRIANGLES

CASE TITLE ---    STRESSES IN A DISC WITH INTERNAL PRESSURE

YOUNGS MODULUS =   .300E+08   POISSONS RATIO =   .300
INITIAL TEMP = 300.0     COEF.OF EXP.=   .500E-05

NODE NO.     X-COORD        Y-COORD
   1       .2000E+01      .0000E+00
   3       .3500E+01      .0000E+00
   5       .5000E+01      .0000E+00
   2       .1932E+01      .5180E+00
   4       .3381E+01      .9060E+00
   6       .4830E+01      .1294E+01

ELEM.NO.     CONNECTING NODES NUMBERED    THICKNESS   TEMP
   1             1       4       2          1.00     300.0
   2             1       3       4          1.00     300.0
   3             3       5       4          1.00     300.0
   4             5       6       4          1.00     300.0

BAND WIDTH =     8

  KNOWN NON-ZERO LOADS
COMPONENT NUMBER    LOAD
        1          .2618E+03

      KNOWN DISPLACEMENTS
COMPONENT NUMBER    DISPLACEMENT
           2          .0000E+00
           6          .0000E+00
          10          .0000E+00

NODE NO.    ALPHA(DEG)   KNOWN U      TANG.FORCE
    2         -75.0      .000E+00      .262E+03
    4         -75.0      .000E+00      .000E+00
    6         -75.0      .000E+00      .000E+00

NODE NO.       FORCE AND DISPLACEMENT COMPONENTS
     1       .2618E+03 -.9857E+03   .9418E-04   .1999E-16
     2       .2356E+00  .1011E+04   .9655E-04   .2587E-04
     3      -.8159E-12 -.7495E+03   .6874E-04   .1211E-16
     4      -.1990E+03  .7426E+03   .6300E-04   .1688E-04
     5      -.1705E-12 -.2534E+03   .5704E-04   .1155E-16
     6      -.6304E+02  .2353E+03   .5640E-04   .1511E-04

ELEM.NO.      SXX          SYY          SXY             THETA     PS1          PS2
   1      -.307E+03    .133E+04   -.201E+03             6.9  -.331E+03    .136E+04
   2      -.375E+03    .447E+03   -.987E+02             6.8  -.387E+03    .458E+03
   3      -.727E+02    .537E+03   -.849E+02             7.8  -.843E+02    .549E+03
   4      -.308E+02    .325E+03   -.607E+02             9.4  -.408E+02    .335E+03
```

Figure 2-11. Output Processed by SP23B for the 4 Element Case

tern about the correct curve. When the number of elements was increased to twelve, which is still an extremely small number, we observe a very close agreement with the theoretical results. It is easily seen that the Finite Element Method would give results which would essentially duplicate the theoretical with a further moderate increase in the number of elements.

Results for Problem 2-2, stress due to temperature variation in a thin disc, are shown in Figure 2-17. Here we observe the same zig-zag scatter of

```
              MAIN SP23B    AUG  20 1971
STRESSES IN A PLATE DUE TO TEMPERATURE AND IN-PLANE LOADS
BY FINITE ELEMENT METHODS USING CONSTANT STRAIN TRIANGLES

CASE TITLE ---    STRESSES IN A DISC WITH INTERNAL PRESSURE

YOUNGS MODULUS =  .300E+08   POISSONS RATIO =  .300
INITIAL TEMP = 300.0     COEF.OF EXP.=  .500E-05

NODE NO.    X-COORD        Y-COORD
    1      .2000E+01     .0000E+00
    3      .2500E+01     .0000E+00
    5      .3000E+01     .0000E+00
    7      .3500E+01     .0000E+00
    9      .4000E+01     .0000E+00
   11      .4500E+01     .0000E+00
   13      .5000E+01     .0000E+00
    2      .1932E+01     .5180E+00
    4      .2415E+01     .6470E+00
    6      .2898E+01     .7760E+00
    8      .3381E+01     .9060E+00
   10      .3864E+01     .1035E+01
   12      .4347E+01     .1165E+01
   14      .4830E+01     .1294E+01

ELEM.NO.      CONNECTING NODES NUMBERED    THICKNESS   TEMP
    1              1        4        2          1.00    300.0
    2              1        3        4          1.00    300.0
    3              3        5        4          1.00    300.0
    4              5        6        4          1.00    300.0
    5              5        8        6          1.00    300.0
    6              5        7        8          1.00    300.0
    7              7        9        8          1.00    300.0
    8              9       10        8          1.00    300.0
    9              9       12       10          1.00    300.0
   10              9       11       12          1.00    300.0
   11             11       13       12          1.00    300.0
   12             13       14       12          1.00    300.0

BAND WIDTH =     8

  KNOWN NON-ZERO LOADS

COMPONENT NUMBER    LOAD
        1          .2618E+03

       KNOWN DISPLACEMENTS
COMPONENT NUMBER    DISPLACEMENT
            2          .0000E+00
            6          .0000E+00
           10          .0000E+00
           14          .0000E+00
           18          .0000E+00
           22          .0000E+00
           26          .0000E+00

NODE NO.   ALPHA(DEG)   KNOWN U      TANG.FORCE
    2       -75.0       .000E+00     .262E+03
    4       -75.0       .000E+00     .000E+00
    6       -75.0       .000E+00     .000E+00
    8       -75.0       .000E+00     .000E+00
   10       -75.0       .000E+00     .000E+00
   12       -75.0       .000E+00     .000E+00
   14       -75.0       .000E+00     .000E+00

NODE NO.       FORCE AND DISPLACEMENT COMPONENTS
    1      .2618E+03 -.3440E+03  .1086E-03  .1492E-16
    2      .1643E+03  .3983E+03  .1075E-03  .2881E-04
```

Figure 2-12. Output Processed by SP23B for the 12-Element Case

3	-.6692E-12	-.4679E+03	.9298E-04	.1137E-16
4	-.1219E+03	.4549E+03	.8813E-04	.2361E-04
5	.6466E-12	-.3571E+03	.8040E-04	.9127E-17
6	-.9227E+02	.3443E+03	.7862E-04	.2107E-04
7	-.1105E-11	-.2871E+03	.7386E-04	.7155E-17
8	-.7407E+02	.2764E+03	.7067E-04	.1894E-04
9	.5258E-12	-.2417E+03	.6829E-04	.5995E-17
10	-.6289E+02	.2347E+03	.6640E-04	.1779E-04
11	-.7523E-12	-.2118E+03	.6530E-04	.5000E-17
12	-.5448E+02	.2033E+03	.6273E-04	.1681E-04
13	.5684E-13	-.7896E+02	.6274E-04	.3909E-17
14	-.2050E+02	.7651E+02	.6081E-04	.1629E-04

ELEM.NO.	SXX	SYY	SXY	THETA	PS1	PS2
1	-.743E+03	.135E+04	-.367E+03	9.7	-.805E+03	.141E+04
2	-.666E+03	.895E+03	-.134E+03	4.9	-.678E+03	.906E+03
3	-.468E+03	.955E+03	-.124E+03	5.0	-.479E+03	.965E+03
4	-.355E+03	.660E+03	-.194E+03	10.5	-.391E+03	.696E+03
5	-.251E+03	.695E+03	-.180E+03	10.4	-.284E+03	.728E+03
6	-.225E+03	.560E+03	-.604E+02	4.4	-.229E+03	.564E+03
7	-.160E+03	.579E+03	-.575E+02	4.4	-.165E+03	.583E+03
8	-.105E+03	.458E+03	-.111E+03	10.8	-.126E+03	.479E+03
9	-.642E+02	.471E+03	-.106E+03	10.8	-.843E+02	.491E+03
10	-.548E+02	.416E+03	-.345E+02	4.2	-.573E+02	.419E+03
11	-.260E+02	.425E+03	-.332E+02	4.2	-.284E+02	.427E+03
12	.504E+01	.362E+03	-.718E+02	10.9	-.884E+01	.376E+03

Figure 2-12 (continued)

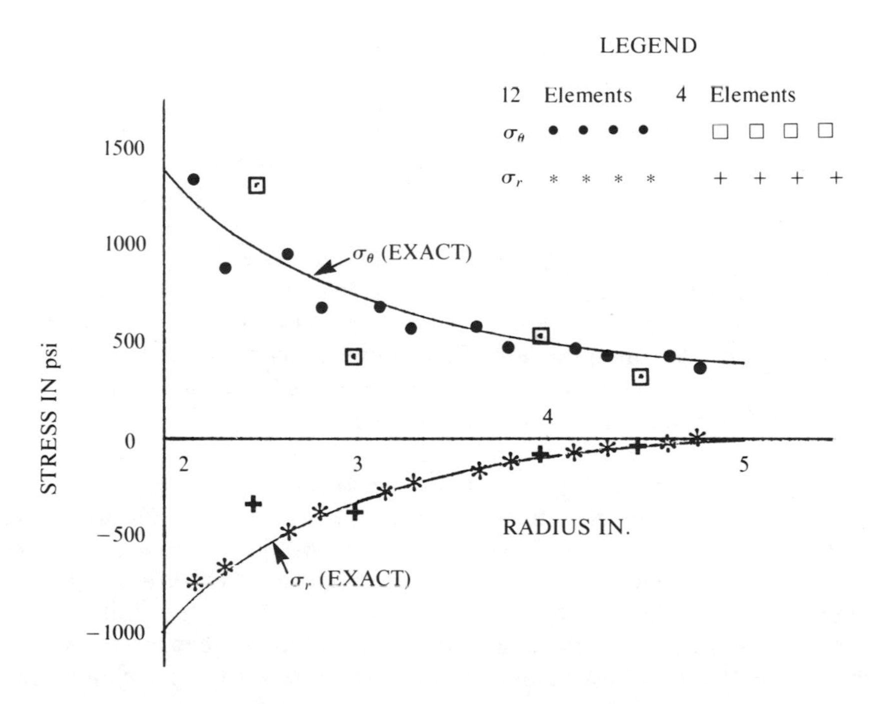

Figure 2-13. Stresses in a Disc with Internal Pressure

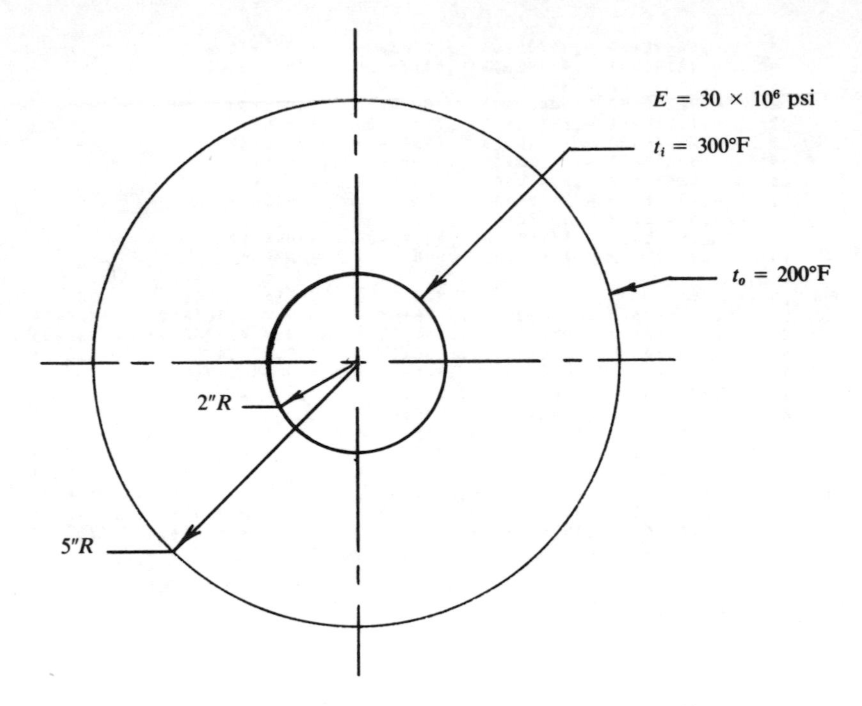

Figure 2-14. Thin Disc with Temperature Gradient

the finite element results about the theoretical curve as was observed in Problem 2-1 (Figure 2-13). Again a small increase in the number of elements from four to twelve gives results that are extremely close to the theoretical results.

In the analysis of a real problem one would use a larger number of elements than were used in these examples. In fact, it is recommended that the analysis be done more than once with varying numbers of elements which will provide an indication of the convergence to the correct answer as the number of elements is increased. In practice, one would terminate the analysis when the variation between two successive sets of results was within the desired accuracy.

Some reference to the interpretation of results was made on page 49, where it was recommended that stresses be taken as being applicable to the material at the centroid of the element. Consequently, edge stresses —which are usually of most interest because they are frequently maximum—must be found by extrapolation. Considering the circumferential stress at the inner surface of Problem 2-1, it can be seen in Figure 2-13 that, if the two innermost element stresses are used to extrapolate to the

```
         MAIN SP23B    AUG  20 1971
STRESSES IN A PLATE DUE TO TEMPERATURE AND IN-PLANE LOADS
BY FINITE ELEMENT METHODS USING CONSTANT STRAIN TRIANGLES

CASE TITLE ---    THERMAL STRESSES IN A THIN DISC

YOUNGS MODULUS =  .300E+08  POISSONS RATIO =  .300
INITIAL TEMP = 300.0    COEF.OF EXP.=  .500E-05

NODE NO.    X-COORD      Y-COORD
    1      .2000E+01    .0000E+00
    3      .3500E+01    .0000E+00
    5      .5000E+01    .0000E+00
    2      .1932E+01    .5180E+00
    4      .3381E+01    .9060E+00
    6      .4830E+01    .1294E+01

ELEM.NO.     CONNECTING NODES NUMBERED   THICKNESS  TEMP
    1            1       4       2          1.00    276.2
    2            1       3       4          1.00    256.6
    3            3       5       4          1.00    225.1
    4            5       6       4          1.00    212.4

BAND WIDTH =     8

  KNOWN NON-ZERO LOADS
COMPONENT NUMBER    LOAD

      KNOWN DISPLACEMENTS
COMPONENT NUMBER    DISPLACEMENT
           2          .0000E+00
           6          .0000E+00
          10          .0000E+00

NODE NO.   ALPHA(DEG)   KNOWN U     TANG.FORCE
    2        -75.0     .000E+00     .000E+00
    4        -75.0     .000E+00     .000E+00
    6        -75.0     .000E+00     .000E+00

NODE NO.       FORCE AND DISPLACEMENT COMPONENTS
    1       .3218E+04  .8540E+04 -.5523E-03 -.1732E-15
    2       .3404E+04 -.7598E+04 -.5773E-03 -.1547E-03
    3       .3060E+04  .1757E+05 -.8938E-03 -.2838E-15
    4       .1039E+05 -.1594E+05 -.8362E-03 -.2241E-03
    5      -.1091E+05  .1020E+05 -.1573E-02 -.4648E-15
    6      -.9160E+04 -.1276E+05 -.1531E-02 -.4103E-03

ELEM.NO.     SXX         SYY         SXY           THETA     PS1        PS2
    1      -.309E+04   -.619E+04   -.443E+03        -8.0 -.302E+04 -.625E+04
    2      -.663E+03   -.112E+04    .389E+03        29.9 -.440E+03 -.134E+04
    3      -.134E+04    .341E+04    .471E+02         -.6 -.134E+04  .341E+04
    4       .320E+02    .348E+04   -.825E+03        12.8 -.156E+03  .366E+04
```

Figure 2-15. Output Processed by SP23B for the 4-Element Case

edge, a significant error is introduced. No careful stress analyst would extrapolate in this way from points plotted as in Figure 2-13, but such an error could be made by interpreting the printed output directly and performing a numerical extrapolation.

It is good practice to plot stresses at points on the critical section and to draw a smooth curve. Stresses zig-zagging above and below the exact curve are typical of finite element solutions, but a smooth curve running

```
            MAIN SP23B    AUG  20 1971
STRESSES IN A PLATE DUE TO TEMPERATURE AND IN-PLANE LOADS
BY FINITE ELEMENT METHODS USING CONSTANT STRAIN TRIANGLES

CASE TITLE ---    THERMAL STRESSES IN A THIN DISC

YOUNGS MODULUS =   .300E+08   POISSONS RATIO =   .300
INITIAL TEMP = 300.0    COEF.OF EXP.=   .500E-05

NODE NO.     X-COORD        Y-COORD
    1      .2000E+01      .0000E+00
    3      .2500E+01      .0000E+00
    5      .3000E+01      .0000E+00
    7      .3500E+01      .0000E+00
    9      .4000E+01      .0000E+00
   11      .4500E+01      .0000E+00
   13      .5000E+01      .0000E+00
    2      .1932E+01      .5180E+00
    4      .2415E+01      .6470E+00
    6      .2898E+01      .7760E+00
    8      .3381E+01      .9060E+00
   10      .3864E+01      .1035E+01
   12      .4347E+01      .1165E+01
   14      .4830E+01      .1294E+01

ELEM.NO.      CONNECTING NODES NUMBERED     THICKNESS    TEMP
    1              1        4         2          1.00     292.1
    2              1        3         4          1.00     284.0
    3              3        5         4          1.00     269.0
    4              5        6         4          1.00     262.8
    5              5        8         6          1.00     250.6
    6              5        7         8          1.00     245.1
    7              7        9         8          1.00     234.6
    8              9       10         8          1.00     229.8
    9              9       12        10          1.00     220.7
   10              9       11        12          1.00     216.5
   11             11       13        12          1.00     208.3
   12             13       14        12          1.00     204.5

BAND WIDTH =      8

  KNOWN NON-ZERO LOADS

COMPONENT NUMBER    LOAD

       KNOWN DISPLACEMENTS
COMPONENT NUMBER     DISPLACEMENT
            2          .0000E+00
            6          .0000E+00
           10          .0000E+00
           14          .0000E+00
           18          .0000E+00
           22          .0000E+00
           26          .0000E+00

NODE NO.    ALPHA(DEG)    KNOWN U      TANG.FORCE
    2         -75.0      .000E+00      .000E+00
    4         -75.0      .000E+00      .000E+00
    6         -75.0      .000E+00      .000E+00
    8         -75.0      .000E+00      .000E+00
   10         -75.0      .000E+00      .000E+00
   12         -75.0      .000E+00      .000E+00
   14         -75.0      .000E+00      .000E+00

NODE NO.          FORCE AND DISPLACEMENT COMPONENTS
    1      .9966E+03   .2717E+04  -.6198E-03  -.1178E-15
    2      .1105E+04  -.2423E+04  -.6192E-03  -.1659E-03
    3      .1040E+04   .4883E+04  -.6491E-03  -.1187E-15
    4      .3234E+04  -.4348E+04  -.6147E-03  -.1647E-03
    5      .1983E+04   .5548E+04  -.7371E-03  -.1418E-15
    6      .2383E+04  -.4973E+04  -.7192E-03  -.1927E-03
```

Figure 2-16. Output Processed by SP23B for the 12-Element Case

```
     7     .1013E+04  .6058E+04 -.8995E-03 -.1510E-15
     8     .3539E+04 -.5450E+04 -.8636E-03 -.2314E-03
     9     .2009E+04  .6855E+04 -.1094E-02 -.1701E-15
    10     .2724E+04 -.6245E+04 -.1060E-02 -.2841E-03
    11     .1016E+04  .7388E+04 -.1331E-02 -.1744E-15
    12     .3889E+04 -.6731E+04 -.1283E-02 -.3437E-03
    13    -.1276E+05  .2455E+04 -.1590E-02 -.1216E-15
    14    -.1217E+05 -.5735E+04 -.1537E-02 -.4119E-03

ELEM.NO.      SXX         SYY          SXY             THETA     PS1        PS2
    1     -.107E+04   -.840E+04    .101E+04             7.7  -.932E+03  -.854E+04
    2     -.103E+04   -.555E+04    .525E+03             6.5  -.974E+03  -.561E+04
    3     -.169E+04   -.350E+04    .348E+03            10.5  -.162E+04  -.356E+04
    4     -.157E+04   -.232E+04    .328E+02             2.5  -.157E+04  -.232E+04
    5     -.161E+04   -.582E+03   -.328E+03            16.3  -.171E+04  -.486E+03
    6     -.146E+04    .135E+03   -.351E+02             1.3  -.147E+04   .136E+03
    7     -.136E+04    .173E+04   -.133E+03             2.5  -.136E+04   .174E+04
    8     -.976E+03    .186E+04   -.631E+03            12.0  -.111E+04   .199E+04
    9     -.727E+03    .325E+04   -.861E+03            11.7  -.905E+03   .343E+04
   10     -.609E+03    .350E+04   -.242E+03             3.4  -.623E+03   .351E+04
   11     -.363E+03    .479E+04   -.311E+03             3.4  -.382E+03   .481E+04
   12      .971E+02    .458E+04   -.948E+03            11.5  -.951E+02   .478E+04
*EXIT*
```

Figure 2-16 (continued)

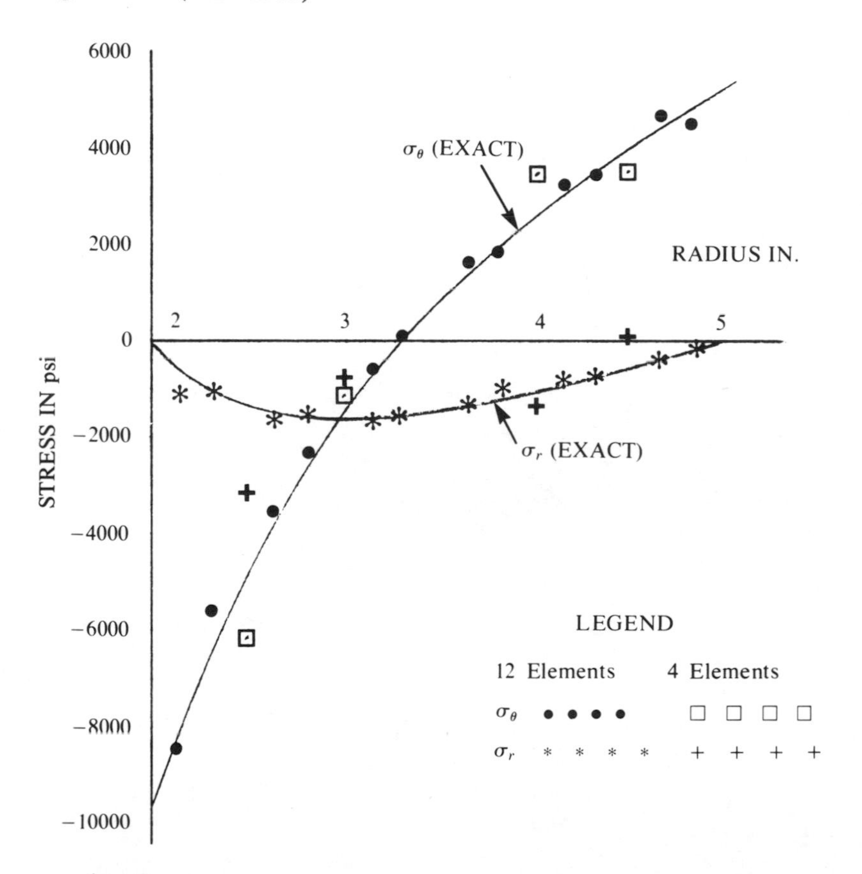

Figure 2-17. Stresses in a Disc with Temperature Variation

between the points usually gives accurate results. If the total load on the critical section is known, it is also good practice to check equilibrium. When there is an imbalance, all stresses on the section can be multiplied by the factor that will establish equilibrium. The stresses so obtained are usually quite accurate.

3 Linear Strain Triangles

Displacement Functions

The constant strain triangle of Chapter 2 provides an introduction that is easily understood. It is also a useful element for practical problems, but does require that rather small elements be used in regions where stress varies abruptly. This means that the number of such elements tends to be large when reliable stress values are needed. By using more advanced elements, in which the stress may vary within the element, much better results can be obtained with the same number of elements. There is no limit to the degree of variation in stress that can be specified and in this chapter only one step in this direction will be taken; provision will be made for the stress to vary linearly within the element.

Later it will be shown that the strain, and hence the stress varies linearly if the displacement functions are of the form

$$u = \alpha_1 + \alpha_2 x + \alpha_3 y + \alpha_4 x^2 + \alpha_5 xy + \alpha_6 y^2$$

$$v = \alpha_7 + \alpha_8 x + \alpha_9 y + \alpha_{10} y^2 + \alpha_{11} xy + \alpha_{12} y^2$$

or,

$$\begin{Bmatrix} u \\ v \end{Bmatrix} = \mathsf{P}\alpha \tag{3.1}$$

where

$$\mathsf{P} = \begin{bmatrix} 1 & x & y & x^2 & xy & y^2 & 0 & 0 & 0 & 0 & 0 & 0 \\ 0 & 0 & 0 & 0 & 0 & 0 & 1 & x & y & x^2 & xy & y^2 \end{bmatrix} \tag{3.2}$$

To determine the six unknown coefficients in the function for u we must know six horizontal displacements at six specified points. The same is true for the unknowns in the v function. This can be accomplished by introducing nodes at the midpoints of the element sides as shown in Figure 3-1a. It will be noted that local axes have been established with origin at the centroid of the element. Dealing with these local axes does not alter the values in the stiffness matrix but does make it easier to perform the integration in the stiffness formula (2.38).

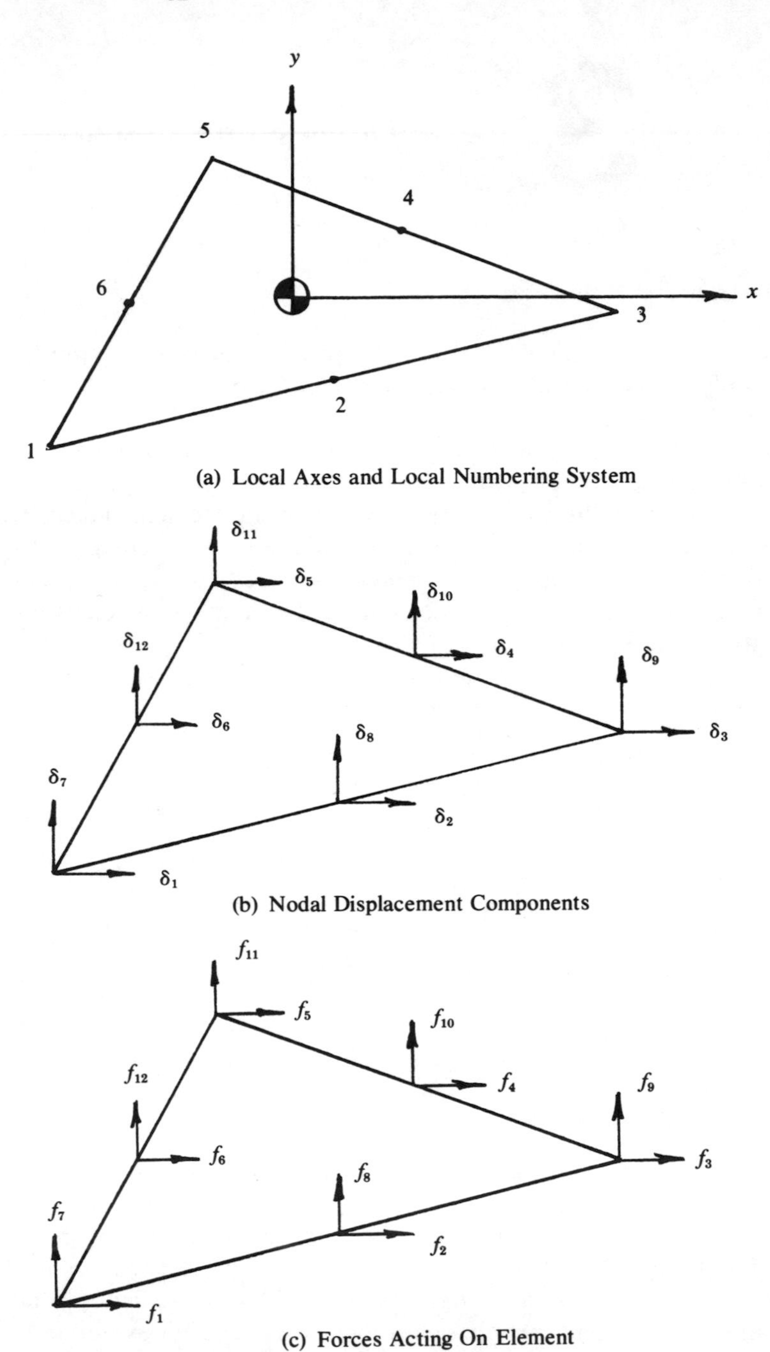

Figure 3-1. Linear Strain Triangle

The displacements at node 1, δ_1 and δ_7, are given by (3.1) and (3.2) as

$$\begin{Bmatrix} \delta_1 \\ \\ \delta_7 \end{Bmatrix} = \mathsf{P}_{\substack{(x=x_1)\\(y=y_1)}} \boldsymbol{\alpha}$$

$$= \begin{bmatrix} 1 & x_1 & y_1 & x_1^2 & x_1y_1 & y_1^2 & 0 & 0 & 0 & 0 & 0 & 0 \\ \\ 0 & 0 & 0 & 0 & 0 & 0 & 1 & x_1 & y_1 & x_1^2 & x_1y_1 & y_1^2 \end{bmatrix} \boldsymbol{\alpha}$$

Similar equations can be written for the δ components at all other nodes. When the δ components are put into numerical order these equations become

$$\boldsymbol{\delta} = \begin{bmatrix} \overline{\mathsf{A}} & 0 \\ \\ 0 & \overline{\mathsf{A}} \end{bmatrix} \boldsymbol{\alpha} \tag{3.3}$$

where

$$\overline{\mathsf{A}} = \begin{bmatrix} 1 & x_1 & y_1 & x_1^2 & x_1y_1 & y_1^2 \\ 1 & x_2 & y_2 & x_2^2 & x_2y_2 & y_2^2 \\ 1 & x_3 & y_3 & x_3^2 & x_3y_3 & y_3^2 \\ 1 & x_4 & y_4 & x_4^2 & x_4y_4 & y_4^2 \\ 1 & x_5 & y_5 & x_5^2 & x_5y_5 & y_5^2 \\ 1 & x_6 & y_6 & x_6^2 & x_6y_6 & y_6^2 \end{bmatrix} \tag{3.4}$$

Equation (3.3) can be written

$$\boldsymbol{\delta} = \mathsf{A}\boldsymbol{\alpha} \tag{3.5}$$

where

$$\mathsf{A} = \begin{bmatrix} \overline{\mathsf{A}} & 0 \\ \\ 0 & \overline{\mathsf{A}} \end{bmatrix} \tag{3.6}$$

Solving (3.5) for $\boldsymbol{\alpha}$ gives

$$\boldsymbol{\alpha} = \mathsf{A}^{-1}\boldsymbol{\delta} \tag{3.7}$$

A is a 12×12 matrix and could be inverted directly, but much time is saved by inverting $\overline{\mathsf{A}}$ and then completing A^{-1} by

$$\mathsf{A}^{-1} = \begin{bmatrix} \overline{\mathsf{A}}^{-1} & 0 \\ \\ 0 & \overline{\mathsf{A}}^{-1} \end{bmatrix} \tag{3.8}$$

The displacement function (3.1) can now be written

$$\begin{Bmatrix} u \\ v \end{Bmatrix} = \mathbf{PA}^{-1}\boldsymbol{\delta} \tag{3.9}$$

We will now check on the compatibility of displacements at points located on the common boundary between adjacent elements. That is, we will determine if pairs of points that touched one another across a boundary before displacement, remain in contact after displacement and hence whether displacement causes gaps to appear as in Figure 2-1b.

Considering only the u displacement, the function in general is

$$u = \alpha_1 + \alpha_2 x + \alpha_3 y + \alpha_4 x^2 + \alpha_5 xy + \alpha_6 y^2$$

However we are interested in u only at points on a particular boundary. Let the boundary be parallel to the y axis which means that all x are constants and the u function takes the parabolic form $u_b = A + By + Cy^2$. Since on this boundary there are three nodes and hence three known values of u_b, constants A, B, and C can be determined and the shape of the displaced edge would be a uniquely determined parabola. If we now consider the u displacements of points just across the boundary, they too vary parabolically and are uniquely determined by the motion of three nodes. Since displacements are common at the three nodal points, the parabolic displacements on both sides of the boundary are identical. Hence there will be no tendancy for gaps to open or for overlap to occur.

A similar treatment of v_b would show that there is no slipping along the interface formed by the boundary. For an oblique boundary the manipulations would be more complex but the conclusions identical. Consequently, displacements are compatible at all element interfaces.

Stiffness of a Linear Strain Triangle

In equation (2.31) the differential operator matrix is

$$\boldsymbol{\Delta} = \begin{bmatrix} \dfrac{\partial}{\partial x} & 0 \\ 0 & \dfrac{\partial}{\partial y} \\ \dfrac{\partial}{\partial y} & \dfrac{\partial}{\partial x} \end{bmatrix}$$

The polynomial matrix has been given in (3.2) and hence from (2.32)

$$\mathsf{B} = \begin{bmatrix} \frac{\partial}{\partial x} & 0 \\ 0 & \frac{\partial}{\partial y} \\ \frac{\partial}{\partial y} & \frac{\partial}{\partial x} \end{bmatrix} \begin{bmatrix} 1 & x & y & x^2 & xy & y^2 & 0 & 0 & 0 & 0 & 0 & 0 \\ 0 & 0 & 0 & 0 & 0 & 0 & 1 & x & y & x^2 & xy & y^2 \end{bmatrix}$$

$$= \begin{bmatrix} 0 & 1 & 0 & 2x & y & 0 & 0 & 0 & 0 & 0 & 0 & 0 \\ 0 & 0 & 0 & 0 & 0 & 0 & 0 & 0 & 1 & 0 & x & 2y \\ 0 & 0 & 1 & 0 & x & 2y & 0 & 1 & 0 & 2x & y & 0 \end{bmatrix} \qquad (3.10)$$

In the strain equation (2.33), the only variables occur in B and these are seen to have a unit exponent. Consequently, a strain varies linearly within the element.

With D given by (2.9) all matrices in the stiffness equation (2.38) are known. The increment of volume dv for an element of uniform thickness, t, can be written

$$dv = t\,dA$$

and (2.38) becomes

$$\mathsf{K} = [\mathsf{A}^{-1}]^T t \int \mathsf{B}^T \mathsf{D} \mathsf{B}\, dA\, \mathsf{A}^{-1} \qquad (3.11)$$

For the constant strain triangle of Chapter 2, the integration to obtain stiffness presented no difficulty since all terms in B and D were constants. In the present case, B contains x and y terms as well as constants. When the product $\mathsf{B}^T\mathsf{D}\mathsf{B}$ is considered, it is evident that we must integrate terms containing constants, x, y, x^2, xy, and y^2. These integrations lead to some intricate formulas except in the case where the origin is at the centroid, then the solutions are quite simple. Even these are difficult to compute when working in Cartesian coordinates; but by means of area coordinates (see [3] and [16]), formulas can be derived.

Using these formulas we find that

$$\int \mathbf{B}^T\mathbf{D}\mathbf{B}\,dA = \frac{\text{area} \times E}{1 - \nu^2} \times$$

$$\begin{bmatrix}
0 & 0 & 0 & 0 & 0 & 0 & 0 & 0 & 0 & 0 & 0 & 0 \\
0 & 1 & 0 & 0 & 0 & 0 & 0 & 0 & \nu & 0 & 0 & 0 \\
0 & 0 & \mu & 0 & 0 & 0 & 0 & \mu & 0 & 0 & 0 & 0 \\
0 & 0 & 0 & 4xx & 2xy & 0 & 0 & 0 & 0 & 0 & 2\nu xx & 4\nu xy \\
0 & 0 & 0 & 2xy & \mu xx+yy & 2\mu xy & 0 & 0 & 0 & 2\mu xx & (\nu+\mu)xy & 2\nu yy \\
0 & 0 & 0 & 0 & 2\mu xy & 4\mu yy & 0 & 0 & 0 & 4\mu xy & 2\mu yy & 0 \\
0 & 0 & 0 & 0 & 0 & 0 & 0 & 0 & 0 & 0 & 0 & 0 \\
0 & 0 & \mu & 0 & 0 & 0 & 0 & \mu & 0 & 0 & 0 & 0 \\
0 & \nu & 0 & 0 & 0 & 0 & 0 & 0 & 1 & 0 & 0 & 0 \\
0 & 0 & 0 & 0 & 2\mu xx & 4\mu xy & 0 & 0 & 0 & 4\mu xx & 2\mu xy & 0 \\
0 & 0 & 0 & 2\nu xx & (\nu+\mu)xy & 2\mu yy & 0 & 0 & 0 & 2\mu xy & xx+\mu yy & 2xy \\
0 & 0 & 0 & 4\nu xy & 2\nu yy & 0 & 0 & 0 & 0 & 0 & 2xy & 4yy
\end{bmatrix} \tag{3.12}$$

where:

$$\mu = (1 - \nu)/2$$

$$xx = 1/12(x_1^2 + x_3^2 + x_5^2)$$

$$xy = 1/12(x_1y_1 + x_3y_3 + x_5y_5)$$

$$yy = 1/12(y_1^2 + y_3^2 + y_5^2)$$

We are now able to evaluate the stiffness matrix for any element using the elastic constants of the material, the element thickness and the coordinates of the nodes based on axes having origin at the centroid of the element. This is done by using equations (3.4), (3.8), (3.12) and (3.11).

Stresses in a Linear Strain Triangle

The stiffness of each element in turn is accumulated as for the cases in Chapter 1 and Chapter 2 to give a stiffness matrix for the assembly of elements. Loads and constraints are applied as before and all displacements, **U**, found. Taking each element in turn the values in $\boldsymbol{\delta}$ can be selected from **U** and used to calculate stress by

$$\boldsymbol{\sigma} = \mathbf{D}\mathbf{B}\mathbf{A}^{-1}\boldsymbol{\delta} \tag{3.13}$$

which is the form that (2.34) takes when the no-load strains are zero.

For the case being considered there are some variable terms in **B** but none in the other matrices of (3.13). The variables all have an exponent equal to unity, hence, the stress varies linearly within the element.

Since stress varies in an element, when solving (3.13) the point at which the stress is to be calculated must be indicated by specifying its location through the x and y values inserted into B.

Results from linear strain triangles (LST) and constant strain triangles (CST), for triangles of equal sizes, when compared with known solutions show that the LST solutions are in general more accurate. However, this depends on the location within the LST element at which stress is determined. Under some conditions, large errors occur; but experience shows that good results are obtained if stresses are calculated at the centroid of the element.

Consistent Load Vector

When concentrated loads occur, they are handled simply by placing a node at the point of load application and taking the load components as known force components. However, a distributed traction acting on an element boundary is more difficult to deal with. In the CST, a traction on an element edge could be treated as concentrated loads at the end nodes by making the two systems statically equivalent. With the LST, there are three nodes on any edge and the principle of static equivalence is not sufficient to determine distribution uniquely. The solution to this problem will be found by resorting to energy principles.

Consider an element boundary such as 1-2-3 in Figure 3-2. If a horizontal displacement occurs, the edge displacement will be parabolic in form and the new edge location may be taken as 1′-2′-3′ as shown by the broken lines. The edge displacement is given by

$$u = ay^2 + by + c = [y^2 \quad y \quad 1] \begin{Bmatrix} a \\ b \\ c \end{Bmatrix}$$

When this displacement takes place, the edge load, p, does work on the system. Over an incremental length of edge dy long the work done is

$$\begin{aligned} dW_p &= put\,dy \\ &= (p_0 + sy)[y^2 \quad y \quad 1] \begin{Bmatrix} a \\ b \\ c \end{Bmatrix} t\,dy \end{aligned}$$

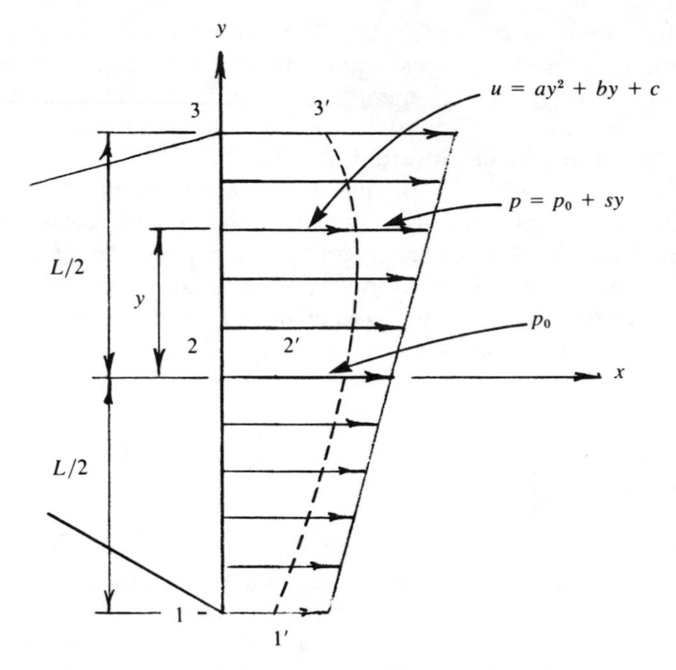

(a) Edge Displacements and Distributed Load

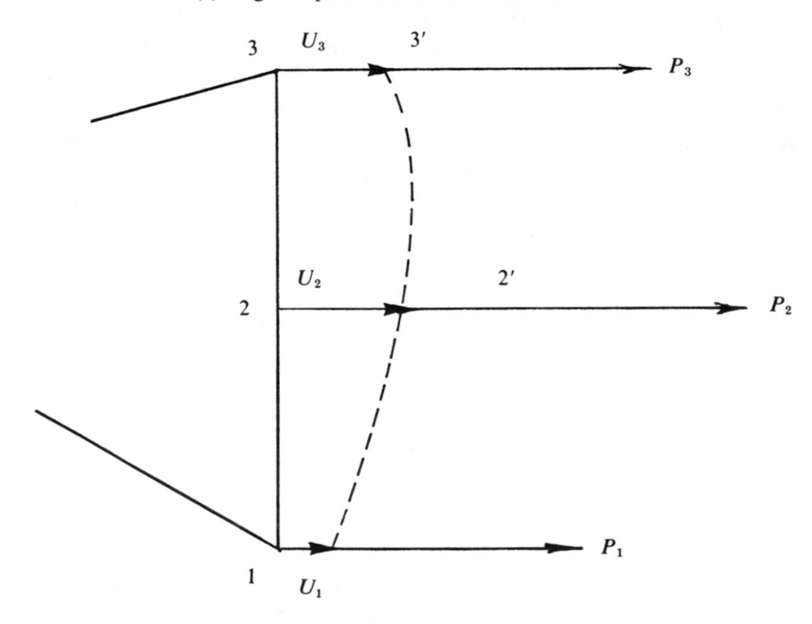

(b) Nodal Displacements and Concentrated Loads

Figure 3-2. Distributed Load and Equivalent Consistent Load Vector

$$= [p_0y^2 + sy^3 \quad p_0y + sy^2 \quad p_0 + sy] \begin{Bmatrix} a \\ b \\ c \end{Bmatrix} t\,dy$$

and the work done on the whole edge is

$$W_p = \int_{-L/2}^{L/2} dW_p$$

$$= t[p_0L^3/12 \quad SL^3/12 \quad p_0L] \begin{Bmatrix} a \\ b \\ c \end{Bmatrix}$$

Consider now a concentrated load system, P_1, P_2 and P_3 which would be used to replace the distributed system. These forces move through distances U_1, U_2, and U_3 which can be written

$$\begin{Bmatrix} U_1 \\ U_2 \\ U_3 \end{Bmatrix} = \begin{Bmatrix} u_{(y=-L/2} \\ u_{(y=0)} \\ u_{(y=L/2)} \end{Bmatrix} = \begin{bmatrix} (-L/2)^2 & -L/2 & 1 \\ 0 & 0 & 1 \\ (L/2)^2 & L/2 & 1 \end{bmatrix} \begin{Bmatrix} a \\ b \\ c \end{Bmatrix}$$

The work done by the concentrated forces is given by

$$W_p = [P_1 \quad P_2 \quad P_3] \begin{Bmatrix} U_1 \\ U_2 \\ U_3 \end{Bmatrix}$$

$$= [P_1 \quad P_2 \quad P_3] \begin{bmatrix} L^2/4 & -L/2 & 1 \\ 0 & 0 & 1 \\ L^2/4 & L/2 & 1 \end{bmatrix} \begin{Bmatrix} a \\ b \\ c \end{Bmatrix}$$

Making the two loading systems equivalent by equating the work done by the systems gives

$$[P_1 \quad P_2 \quad P_3] \begin{bmatrix} L^2/4 & -L/2 & 1 \\ 0 & 0 & 1 \\ L^2/4 & L/2 & 1 \end{bmatrix} \begin{Bmatrix} a \\ b \\ c \end{Bmatrix}$$

$$= t[p_0L^3/12 \quad SL^3/12 \quad p_0L] \begin{Bmatrix} a \\ b \\ b \end{Bmatrix}$$

When the right-hand element is removed from both sides and both sides are post multiplied by the inverse of

$$\begin{bmatrix} L^2/4 & -L/2 & 1 \\ 0 & 0 & 1 \\ L^2/4 & L/2 & 1 \end{bmatrix}$$

which is

$$\begin{bmatrix} 2/L^2 & -4/L^2 & 2/L^2 \\ -1/L & 0 & 1/L \\ 0 & 1 & 0 \end{bmatrix}$$

we get

$$[P_1 \quad P_2 \quad P_3] = t[p_0L^3/12 \quad SL^3/12 \quad p_0L] \begin{bmatrix} 2/L^2 & -4/L^2 & 2/L^2 \\ -1/L & 0 & 1/L \\ 0 & 1 & 0 \end{bmatrix}$$

$$= t[p_0L/6 - SL^2/12 \quad -p_0L/3 + p_0L \quad p_0L/6 + SL^2/12]$$

Hence,

$$P_1 = (tL/6)(p_0 - SL/2)$$
$$P_2 = (2tL/3)p_0$$
$$P_3 = (tL/6)(p_0 + SL/2)$$

Using p_1, p_2, and p_3 to represent the intensity of the distributed load at points 1, 2, and 3, and noting that tL is the edge area, we get

$$\begin{aligned} P_1 &= (1/6)p_1 \times \text{edge area} \\ P_2 &= (2/3)p_2 \times \text{edge area} \\ P_3 &= (1/6)p_3 \times \text{edge area} \end{aligned} \tag{3.14}$$

These are the components of the consistent load vector. They are statically equivalent to the distributed load and also do the same amount of work during a displacement. The work equality was based on an edge displacement function which is consistent with the assumed displacement function for all points in the element, hence, the designation ''consistant load vector.''

A Program to Determine Stresses in a Plate by Linear Strain Triangles

Program SP33B uses linear strain triangles to solve in-plane loaded plate problems. The user must subdivide the plate to be analyzed into triangular elements and number the nodes at all element corners and at all midpoints on the sides. To distinguish between the two types of nodes, those at corners of triangles will be referred to as *principal nodes* and those on the sides as *secondary nodes*.

While reading the following, frequent reference should be made to Figure 3-3, SP33B Flow Chart; Figure 3-4, SP33B Fortran Statements; Figure 3-5, Instructions for SP33B Data Deck Preparation; and the computer output as given in figures following the example. The main title, case title, and elastic constants are handled as in SP23B. Following the elastic properties data card, a card is read which contains the number of nodes in the system. It is followed by node data cards for the principal nodes only, none being required for the secondary nodes. The node data cards have the usual format and they are terminated by a blank card as usual.

The element data cards follow and contain the element number, the thickness, and the six node numbers on the boundary. The node numbers must start at a principal node and be in the order found by proceeding counterclockwise around the element. It will be noted that the coordinates of secondary nodes have not been read; the computer determines these by taking the mean of the values that apply to the principal nodes at the ends of the side on which the secondary node is located.

As each element is treated, the coordinates are recalculated for a local system with origin at the centroid of the element. It is these local coordinates that are used in A and in all subsequent references to points in the element. The stiffness of each element is calculated as described earlier and the general stiffness matrix formed by accumulation.

The known forces are read from cards in the usual manner. Concentrated loads are entered directly; but where there are distributed loads, the user must first calculate the consistent load vector components from (3.14) and, if the boundary is inclined, resolve these into x and y components.

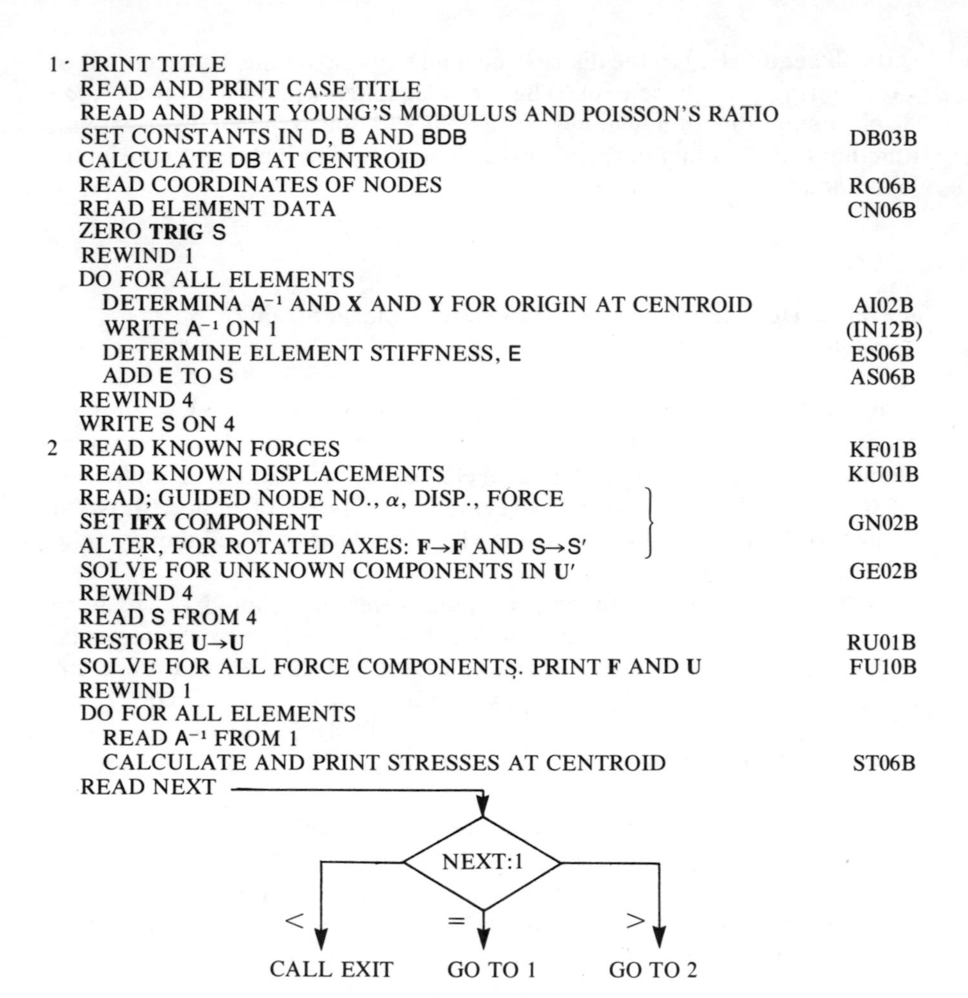

Figure 3-3. SP33B Flow Chart

Known displacements and guided nodes are treated as before and the unknowns are solved as previously. In determining stresses in the elements, (3.13) is used with the values of x and y in $\mathbf{B}$ taken as zero. This gives the stresses at the centroid of the element without offering the user any option as to location. The reason for this restriction is that stresses at other points can be very misleading.

When a case has been completed, the user has three options; end the run, start an entirely new case, or solve the same plate with a new set of loads and constraints.

```
C            *****      MAIN SP33B      *****          JAN 18,1973
C   STRESSES IN A PLATE USING LINEAR STRAIN TRIANGLES      W H BOWES
C       GUIDED NODES
        DIMENSION S(42,200),F(200),U(200),IFX(200),XY(2,100)
        DIMENSION TRIG(200)
        DIMENSION NCON(6,50),T(50)
        DIMENSION E(12,12),B(3,12),AI(12,12),X(3),Y(3),NAME(20)
        DIMENSION DB(3,12),D(3,3),BDB(12,12)
        LBAND=42
      1 WRITE(3,102)
        READ(2,100)NAME
        WRITE(3,101)NAME
        READ(2,103)YM,GNU
        WRITE(3,104)YM,GNU
        CALL DB03B(YM,GNU,D,B,BDB)
        DO 4 I=1,3
        DO 4 J=1,12
        DBIJ=0.
        DO 3 IJ=1,3
      3 DBIJ=DBIJ+D(I,IJ)*B(IJ,J)
      4 DB(I,J)=DBIJ
        CALL RC06B(2,XY,NDF)
        CALL  CN06B(LBAND,6,2,NCON,T,NE,NBAND)
        DO 5 J=1,NDF
        TRIG(J)=0.
        DO 5 I=1,NBAND
      5 S(I,J)=0.
        REWIND 1
        DO 25 IE=1,NE
        CALL AI02B(IE,NCON,XY,T,AI,X,Y)
        WRITE(1) AI
        V=T(IE)*.5*(X(1)*(Y(2)-Y(3))+X(2)*(Y(3)-Y(1))+X(3)*(Y(1)-Y(2)))
        CALL ES06B(V,AI,BDB,X,Y,E)
     25 CALL AS06B(IE,NCON,E,S,2,6,12,LBAND)
        END FILE 1
        REWIND 4
        DO 26 I=1,NBAND
     26 WRITE(4)(S(I,J),J=1,NDF)
        END FILE 4
      2 CALL KF01B(F,NDF)
        CALL KU01B(U,IFX,NDF)
        CALL GN02B(S,J,F,IFX,NBAND,LBAND,TRIG)
        CALL GE02B(F,J,S,IFX,NBAND,NDF,LBAND)
        REWIND 4
        DO 30 I=1,NBAND
     30 READ(4)(S(I,J),J=1,NDF)
        CALL RU01B(U,TRIG,NDF)
        CALL FU10B(S,J,NBAND,NDF,2,LBAND)
        REWIND 1
        WRITE(3,106)
        DO 45 IE=1,NE
        READ(1) AI
     45 CALL ST06B(DB,AI,U,NCON,IE)
        READ(2,108)NEXT
        IF(NEXT-1)47,1,2
     47 CALL EXIT
    100 FORMAT(20A4)
    101 FORMAT('0CASE TITLE  --- ',20A4)
    102 FORMAT('1               MAIN SP33B    JAN 18,1973',/,' STRESSES IN IN-P
       1LANE LOADED PLATE USING LINEAR STRAIN TRIANGLES  (GUIDED NODES)')
    103 FORMAT(2F10.5)
    104 FORMAT('0YOUNGS MODULUS=',E10.3,'  POISSONS RATIO=',F6.3)
    106 FORMAT('0ELEM.NO.     SXX        SYY        SXY       THETA      PS1
       1      PS2')
    108 FORMAT(I5)
        END
```

Figure 3-4. SP33B Fortran Statements

Data Deck:

A card containing the Case Title.

A card containing physical properties;

Young's Modulus and Poisson's Ratio. Format: 2F10.5

A card containing the number of nodes in the system. Format: I5

One card for each node that is located at the corner of a triangle, containing;

Node number, x coordinate, y coordinate. Format: I5, 2F10.5

A blank card to indicate end of node data.

One card for each element containing;

Element number, Thickness of element (=1 by default), Node numbers on sides of triangle, in order, starting at a corner node. Format: I5, F10.5, 6I5

A blank card to indicate end of element data.

One card for each known, nonzero load component containing;

Component number (twice node number minus one for x component or twice node number for y component), Magnitude of force.
Format: I5, F10.5

A blank card to indicate end of load data.

One card for each known displacement component, containing;

Component number, Displacement. Format: I5, F10.5
Note: Zero displacement must be entered.

A blank card to indicate end of displacement data.

One card for each guided node, containing:

Node number, Angle (alpha) in degrees defining direction of guiding plane, Normal displacement of guiding plane, Load tangential to guiding plane. Format: I5, 3F10.5
Note: Consider a set of axes, x′ and y′, rotated so that x′ is normal to the guiding plane. The angle of the x′ axis when measured counterclockwise from the x axis gives alpha. Normal displacement is positive if the plane moves in the +x′ direction. The load is positive if it is in the +y′ direction.

A blank card to indicate end of guided node data.

A card containing NEXT. Format I5

NEXT = 0; End of job
NEXT = 1; Execute program again starting a new case. Follow by an entire deck prepared in accordance with the above instructions starting with the Case Title.
NEXT = 2; Repeat the case just completed but with a new set of known loads and displacements. Follow by force data cards and all following cards described above.

Figure 3-5. Instructions for SP33B Data Deck Preparation

Example

The examples used in Chapter 2 show the nature of the output and the comparison of the finite element results for different numbers of elements with the theoretical stress values for a thin disc with internal pressure as well as a variation in temperature. As has been mentioned earlier, the finite element results are expected to be more accurate when the linear strain triangle is used as compared to the results found using the constant strain

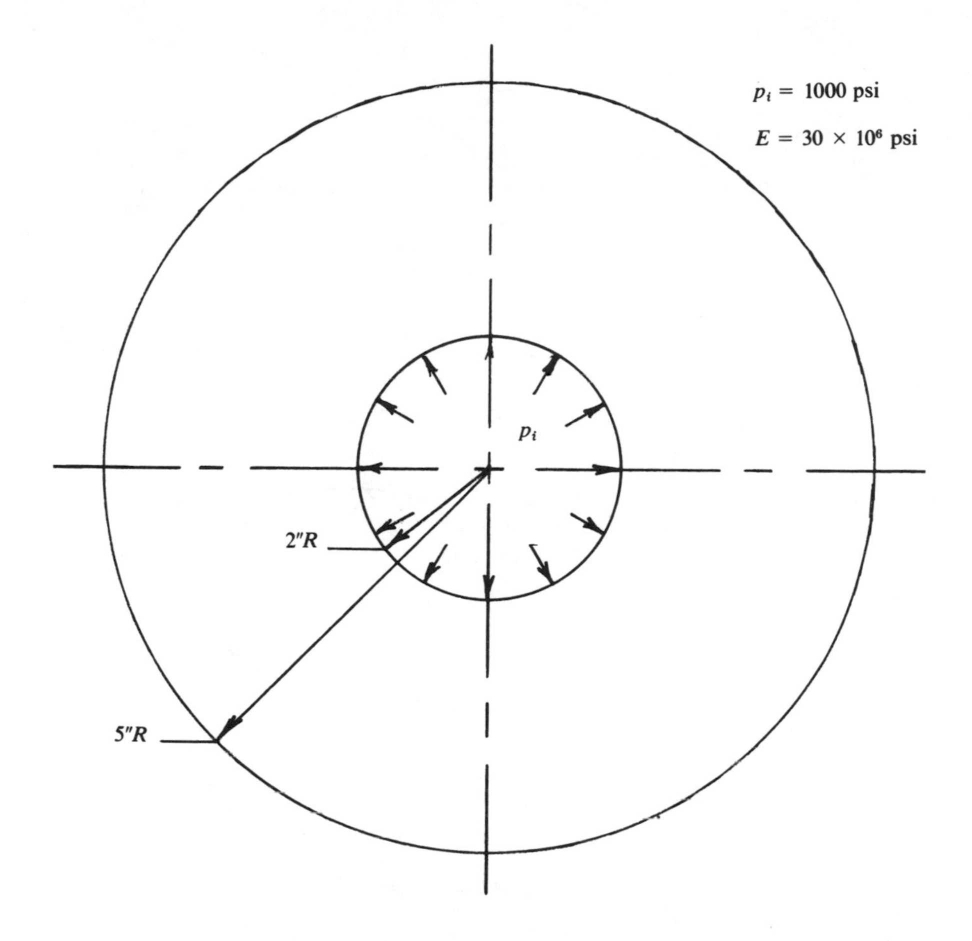

Figure 3-6. Thin Disc With Internal Pressure

triangle. For this reason, the example for this chapter will be the same as that of Chapter 2. This will enable a comparison of the linear strain triangle and the constant strain triangle and the theoretical results to be made.

Problem

For the thin disc shown in Figure 3-6, determine the radial and circumferential stresses by the Finite Element Method, using a linear strain triangle, and compare the results with those found using the constant strain triangle of Chapter 2 and with the theoretical results.

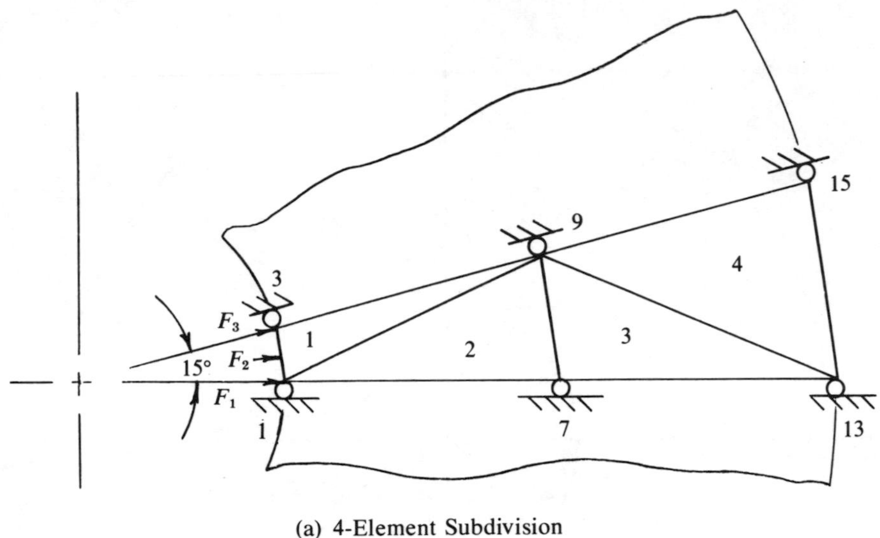

(a) 4-Element Subdivision

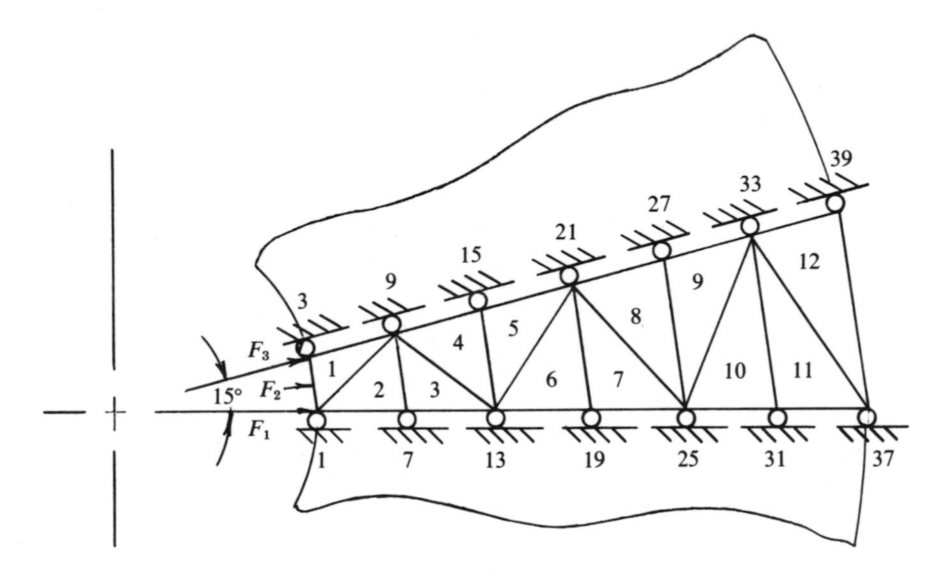

(b) 12-Element Subdivision

Figure 3-7. Element Subdivision for Thin Disc (a) 4-Element Subdivision (b) 12-Element Subdivision

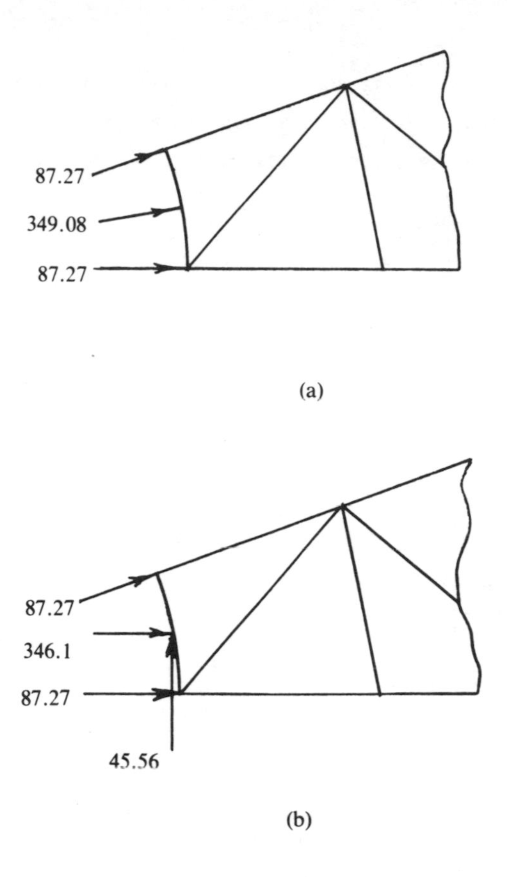

Figure 3-8. Consistent Load Vector

Solution. In order to permit a comparison to be made, the subdivision of the section of the disc will be chosen identical to the element subdivision used for the example in Chapter 2 and as shown in Figure 3-7.

The internal pressure loading is represented for the Finite Element Method by equivalent nodal point loads. Unlike the equivalent nodal point loads as found for the constant strain triangle, the consistent load vector must be chosen as detailed on pages 67-71. The consistent nodal point loads as calculated using equation 3.14 are shown in Figure 3-8a. The x and y components, excepting for the load tangential to the guiding plane, are fed into the computer in the normal manner. The actual components read into the computer are shown in Figure 3-8b.

The boundary conditions are determined from the conditions of symmetry and are shown in Figure 3-7.

```
              MAIN SP33B    JAN 18,1973
STRESSES IN IN-PLANE LOADED PLATE USING LINEAR STRAIN TRIANGLES  (GUIDED NODES

CASE TITLE  ---  STRESSES IN A DISC WITH INTERNAL PRESSURE

YOUNGS MODULUS=  .300E+08  POISSONS RATIO=  .300

NODE NO.    X-COORD      Y-COORD
    1      .2000E+01    .0000E+00
    7      .3500E+01    .0000E+00
   13      .5000E+01    .0000E+00
    3      .1932E+01    .5180E+00
    9      .3381E+01    .9060E+00
   15      .4830E+01    .1294E+01

SYSTEM HAS  30 DEGREES OF FREEDOM

ELEM.NO.   THICKNESS     CONNECTING NODE NUMBERS
    1         1.00         1   5   9   6   3   2
    2         1.00         1   4   7   8   9   5
    3         1.00         7  10  13  11   9   8
    4         1.00        13  14  15  12   9  11

BAND WIDTH =    18

   KNOWN NON-ZERO LOADS
COMPONENT NUMBER    LOAD
         1        .8727E+02
         3        .3461E+03
         4        .4556E+02

       KNOWN DISPLACEMENTS
COMPONENT NUMBER    DISPLACEMENT
          2          .0000E+00
          8          .0000E+00
         14          .0000E+00
         20          .0000E+00
         26          .0000E+00

NODE NO.   ALPHA(DEG)   KNOWN U     TANG.FORCE
   3        -75.0      .000E+00     .873E+02
   6        -75.0      .000E+00     .000E+00
   9        -75.0      .000E+00     .000E+00
  12        -75.0      .000E+00     .000E+00
  15        -75.0      .000E+00     .000E+00

NODE NO.       FORCE AND DISPLACEMENT COMPONENTS
    1      .8727E+02 -.3649E+03  .1101E-03  .7399E-17
    2      .3461E+03  .4556E+02  .1107E-03  .1484E-04
    3     -.9765E+01  .3736E+03  .1081E-03  .2898E-04
    4     -.4187E-12 -.8046E+03  .8710E-04  .1048E-16
    5     -.9095E-12 -.2274E-12  .8596E-04  .1437E-04
    6     -.2105E+03  .7855E+03  .8341E-04  .2235E-04
    7     -.2593E-11 -.2978E+03  .7387E-04  .4811E-17
    8     -.2793E-11 -.5450E-13  .7372E-04  .9660E-05
    9     -.7406E+02  .2764E+03  .7160E-04  .1919E-04
   10     -.1451E-11 -.4484E+03  .6689E-04  .5079E-17
   11     -.1167E-11  .1303E-12  .6691E-04  .7233E-05
   12     -.1171E+03  .4370E+03  .6490E-04  .1739E-04
   13     -.1137E-11 -.8427E+02  .6325E-04  .3842E-17
   14     -.1093E-11 -.8101E-13  .6279E-04  .8287E-05
   15     -.2195E+02  .8193E+02  .6087E-04  .1631E-04

ELEM.NO.     SXX        SYY        SXY         THETA     PS1        PS2
    1     -.555E+03  .959E+03 -.280E+03      10.1 -.605E+03  .101E+04
    2     -.351E+03  .724E+03 -.106E+03       5.6 -.361E+03  .735E+03
    3     -.105E+03  .491E+03 -.470E+02       4.5 -.109E+03  .495E+03
    4     -.359E+02  .415E+03 -.793E+02       9.7 -.494E+02  .429E+03
```

Figure 3-9. Output Processed by the SP33B for the 4-Element Case

```
          MAIN SP33B    JAN 18,1973
STRESSES IN IN-PLANE LOADED PLATE USING LINEAR STRAIN TRIANGLES  (GUIDED NODES)

CASE TITLE  ---  STRESSES IN A DISC WITH INTERNAL PRESSURE

YOUNGS MODULUS=  .300E+08  POISSONS RATIO=  .300

NODE NO.    X-COORD      Y-COORD
    1      .2000E+01    .0000E+00
    7      .2500E+01    .0000E+00
   13      .3000E+01    .0000E+00
   19      .3500E+01    .0000E+00
   25      .4000E+01    .0000E+00
   31      .4500E+01    .0000E+00
   37      .5000E+01    .0000E+00
    3      .1932E+01    .5180E+00
    9      .2415E+01    .6470E+00
   15      .2898E+01    .7760E+00
   21      .3381E+01    .9060E+00
   27      .3864E+01    .1035E+01
   33      .4347E+01    .1165E+01
   39      .4830E+01    .1294E+01

SYSTEM HAS  78 DEGREES OF FREEDOM

ELEM.NO.   THICKNESS     CONNECTING NODE NUMBERS
    1         1.00         1    5    9    6    3    2
    2         1.00         1    4    7    8    9    5
    3         1.00         7   10   13   11    9    8
    4         1.00        13   14   15   12    9   11
    5         1.00        13   17   21   18   15   14
    6         1.00        13   16   19   20   21   17
    7         1.00        19   22   25   23   21   20
    8         1.00        25   26   27   24   21   23
    9         1.00        25   29   33   30   27   26
   10         1.00        25   28   31   32   33   29
   11         1.00        31   34   37   35   33   32
   12         1.00        37   38   39   36   33   35

BAND WIDTH =    18

  KNOWN NON-ZERO LOADS

  COMPONENT NUMBER     LOAD
          1          .8727E+02
          3          .3461E+03
          4          .4556E+02

       KNOWN DISPLACEMENTS
  COMPONENT NUMBER    DISPLACEMENT
              2        .0000E+00
              8        .0000E+00
             14        .0000E+00
             20        .0000E+00
             26        .0000E+00
             32        .0000E+00
             38        .0000E+00
             44        .0000E+00
             50        .0000E+00
             56        .0000E+00
             62        .0000E+00
             68        .0000E+00
             74        .0000E+00

  NODE NO.   ALPHA(DEG)    KNOWN U      TANG.FORCE
     3         -75.0      .000E+00      .873E+02
     6         -75.0      .000E+00      .000E+00
     9         -75.0      .000E+00      .000E+00
    12         -75.0      .000E+00      .000E+00
```

Figure 3-10. Output Processed by the SP33B for the 12-Element Case

```
 15       -75,0       ,000E+00     ,000E+00
 18       -75,0       ,000E+00     ,000E+00
 21       -75,0       ,000E+00     ,000E+00
 24       -75,0       ,000E+00     ,000E+00
 27       -75,0       ,000E+00     ,000E+00
 30       -75,0       ,000E+00     ,000E+00
 33       -75,0       ,000E+00     ,000E+00
 36       -75,0       ,000E+00     ,000E+00
 39       -75,0       ,000E+00     ,000E+00

NODE NO,       FORCE AND DISPLACEMENT COMPONENTS
    1      ,8727E+02 -,1293E+03  ,1115E-03  ,5610E-17
    2      ,3461E+03  ,4556E+02  ,1112E-03  ,1477E-04
    3      ,5054E+02  ,1486E+03  ,1079E-03  ,2892E-04
    4     -,1484E-11 -,3748E+03  ,1013E-03  ,7635E-17
    5     -,1080E-11  ,5116E-12  ,1008E-03  ,1476E-04
    6     -,9728E+02  ,3630E+03  ,9782E-04  ,2621E-04
    7     -,2886E-11 -,1615E+03  ,9321E-04  ,3924E-17
    8     -,1497E-11  ,3649E-12  ,9302E-04  ,1215E-04
    9     -,4052E+02  ,1512E+03  ,9010E-04  ,2414E-04
   10     -,4502E-12 -,2719E+03  ,8686E-04  ,4484E-17
   11      ,7501E-12 -,5504E-12  ,8679E-04  ,1037E-04
   12     -,7048E+02  ,2630E+03  ,8398E-04  ,2250E-04
   13     -,1847E-11 -,1180E+03  ,8180E-04  ,3016E-17
   14     -,1120E-11 -,2999E-13  ,8153E-04  ,1080E-04
   15     -,3166E+02  ,1182E+03  ,7900E-04  ,2117E-04
   16     -,2722E-12 -,2130E+03  ,7766E-04  ,4462E-17
   17     -,1762E-11  ,5116E-12  ,7727E-04  ,1097E-04
   18     -,5514E+02  ,2058E+03  ,7499E-04  ,2009E-04
   19     -,2507E-11 -,9831E+02  ,7425E-04  ,2450E-17
   20     -,3274E-12 -,1651E-12  ,7397E-04  ,9686E-05
   21     -,2454E+02  ,9158E+02  ,7173E-04  ,1922E-04
   22     -,1115E-11 -,1758E+03  ,7146E-04  ,2966E-17
   23      ,1402E-12 -,1310E-12  ,7124E-04  ,8746E-05
   24     -,4553E+02  ,1699E+03  ,6903E-04  ,1850E-04
   25      ,1030E-12 -,8007E+02  ,6917E-04  ,1986E-17
   26     -,2757E-11 -,1951E-12  ,6883E-04  ,9100E-05
   27     -,2148E+02  ,8017E+02  ,6679E-04  ,1790E-04
   28     -,1108E-11 -,1511E+03  ,6726E-04  ,2981E-17
   29     -,2522E-12 -,5684E-13  ,6686E-04  ,9330E-05
   30     -,3912E+02  ,1460E+03  ,6495E-04  ,1740E-04
   31     -,1654E-11 -,7227E+02  ,6568E-04  ,1706E-17
   32     -,3809E-11  ,1035E-12  ,6535E-04  ,8576E-05
   33     -,1812E+02  ,6764E+02  ,6344E-04  ,1700E-04
   34     -,1171E-11 -,1338E+03  ,6439E-04  ,2149E-17
   35      ,2469E-11 -,1145E-12  ,6411E-04  ,7992E-05
   36     -,3466E+02  ,1294E+03  ,6218E-04  ,1666E-04
   37     -,2274E-12 -,2014E+02  ,6333E-04  ,9971E-18
   38     -,1061E-11 -,2918E-13  ,6300E-04  ,8314E-05
   39     -,5354E+01  ,1998E+02  ,6114E-04  ,1638E-04

ELEM,NO.    SXX        SYY        SXY           THETA    PS1         PS2
    1     -,767E+03  ,115E+04 -,352E+03        10,1 -,830E+03   ,121E+04
    2     -,673E+03  ,106E+04 -,160E+03         5,2 -,688E+03   ,107E+04
    3     -,470E+03  ,855E+03 -,111E+03         4,8 -,480E+03   ,864E+03
    4     -,371E+03  ,753E+03 -,199E+03         9,7 -,405E+03   ,787E+03
    5     -,254E+03  ,638E+03 -,166E+03        10,2 -,284E+03   ,668E+03
    6     -,231E+03  ,614E+03 -,779E+02         5,2 -,238E+03   ,621E+03
    7     -,158E+03  ,541E+03 -,594E+02         4,8 -,163E+03   ,546E+03
    8     -,114E+03  ,497E+03 -,109E+03         9,8 -,133E+03   ,516E+03
    9     -,656E+02  ,449E+03 -,956E+02        10,2 -,828E+02   ,466E+03
   10     -,579E+02  ,441E+03 -,456E+02         5,2 -,621E+02   ,445E+03
   11     -,239E+02  ,407E+03 -,364E+02         4,8 -,270E+02   ,410E+03
   12     -,983E-01  ,383E+03 -,693E+02         9,9 -,123E+02   ,395E+03
*EXIT*
```

Figure 3-10 (continued)

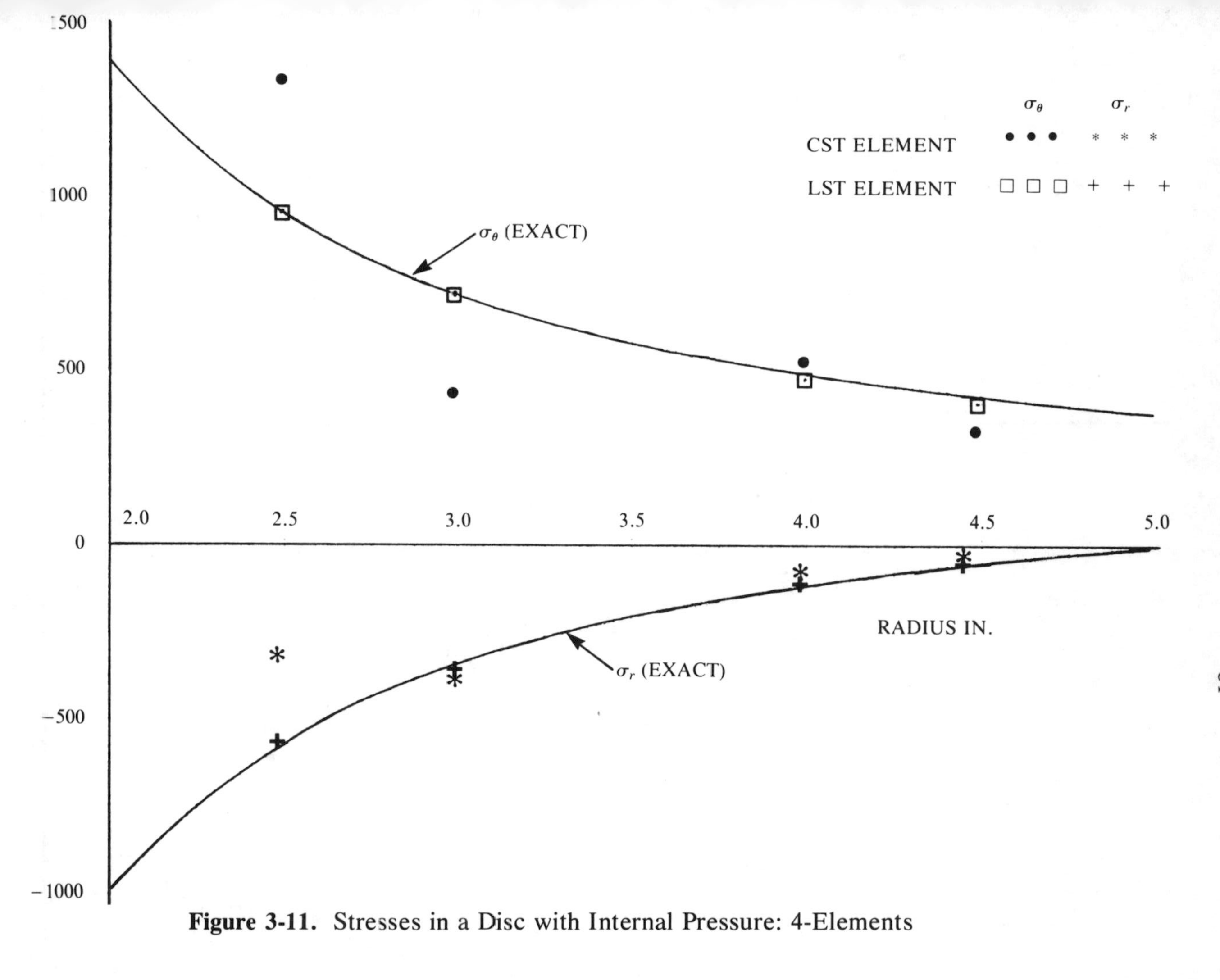

Figure 3-11. Stresses in a Disc with Internal Pressure: 4-Elements

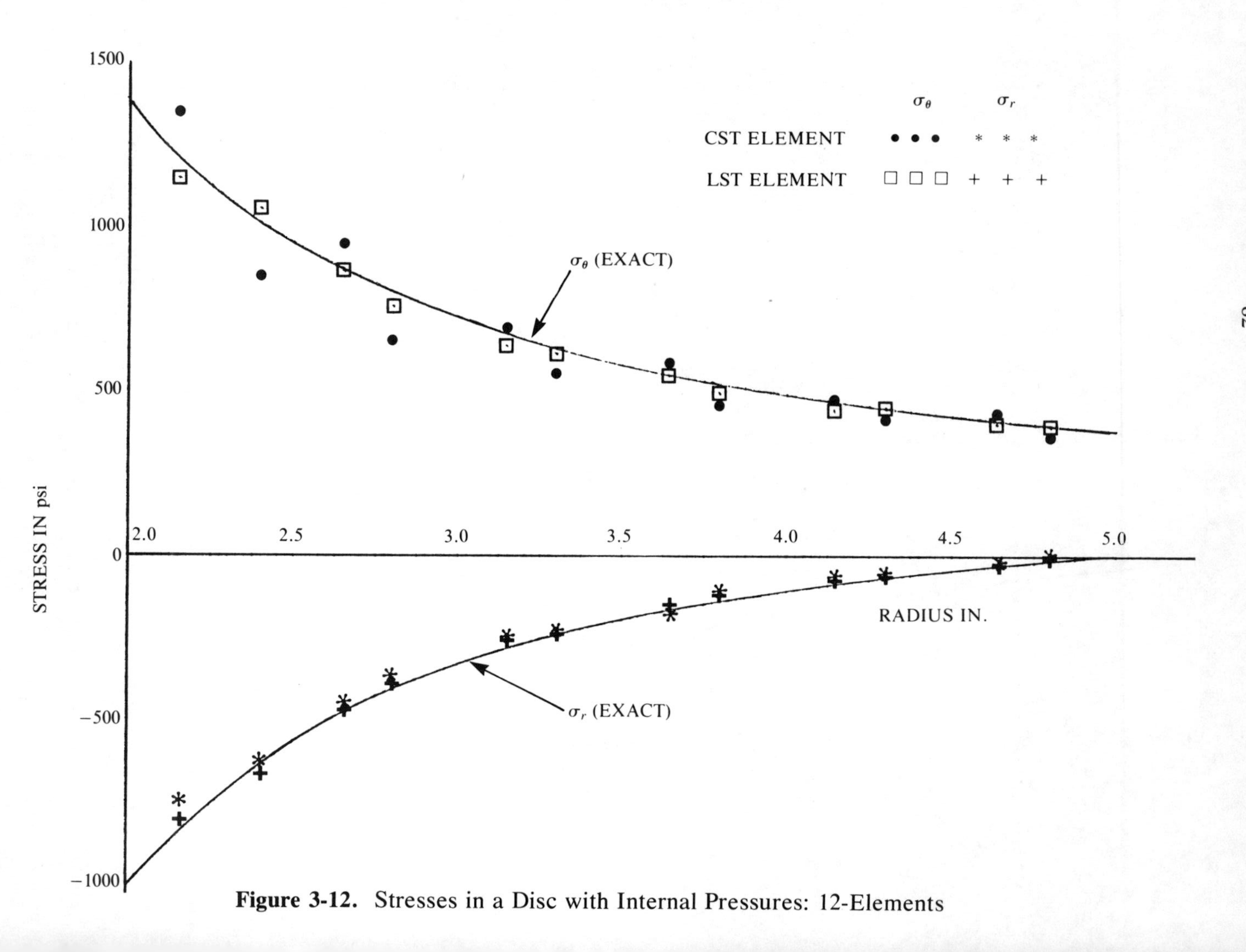

Figure 3-12. Stresses in a Disc with Internal Pressures: 12-Elements

Results

The computer output for this example is shown and interpreted graphically in the figures that follow.

In Figure 3-11, the results using the linear strain triangle of this chapter and the results using the constant strain triangle of Chapter 2 are compared with the theoretical solution. Using only four elements, good accuracy is obtained with the linear strain triangle; a great improvement over the results using the constant strain triangle. It can easily be seen that linear strain elements will normally produce much more accurate results than a larger number of constant strain elements. This can be seen by comparing Figure 3-11 and Figure 3-12. Here we observe that the accuracy obtained using four linear strain triangles is superior to that obtained using twelve constant strain triangles. We observed in Chapter 2, Figure 2-13, and here in Figure 3-12 that good accuracy was obtained with twelve constant strain triangles; but, we also observe in Figure 3-12 that a substantial improvement is obtained with the linear strain triangle, especially in the region where the stress is varying most rapidly.

As explained in Chapter 2, it is always good practice to plot the results and draw a smooth curve through them. If this is done for this problem, the stresses found at the boundary would be very close to the theoretical values.

4 Reinforced Linear Strain Triangles

It is possible to use the methods and programs presented in Chapters 2 and 3 to solve problems that are not strictly plate problems. For example, if an analysis of the I beam of Figure 4-1a is required, it can be treated by considering the beam to be a plate composed of the triangular elements of Figure 4-1b, with those in the flange area having a thickness equal to the width of the flange and those in the web area having a thickness equal to the web thickness. Results using constant strain triangles would prove unsatisfactory for the subdivision shown, while those from linear strain triangles would be more reliable. As with any solution, there will be some loss of accuracy due to numerical round-off during computation. This is discussed by Rosanoff and Ginsberg [11]. For the proportions shown in Figure 4-1b, the error would not be great enough to seriously alter the stresses. If, however, the flanges were very thin or if the length of the beam were very large, the loss of accuracy would have to be seriously considered. For this reason, it is recommended that double precision be used in all calculations.

Similarly, the problem in Figure 4-2a can be solved by triangular elements as indicated in Figure 4-2b. In this case, the thickness of the elements that are substituted for the reinforcement bars must be adjusted to take into account the cross-sectional area of the bars and the difference in Young's modulus. The stresses that are calculated for these elements require some interpretation.

While good results can be obtained in the manner suggested, a great many degrees of freedom are wasted when triangles are used for portions of the structure that function essentially as bar elements. A far better method is to treat the structure as though it is a composite of two distinct families of elements: triangular plates and bars.

Stiffness Matrix for a Composite Structure

The problems of Figures 4-1a and 4-2a can be treated as plates reinforced by bars as shown in Figures 4-1c and 4-2c. There are some approximations involved in this solution. For example, the flanges are assumed to be uniformly stressed over their thickness and over the length between neighbouring nodes. This approximation can be very good if the row of bars is placed at the centroidal axis of the flange and if bar elements are kept short

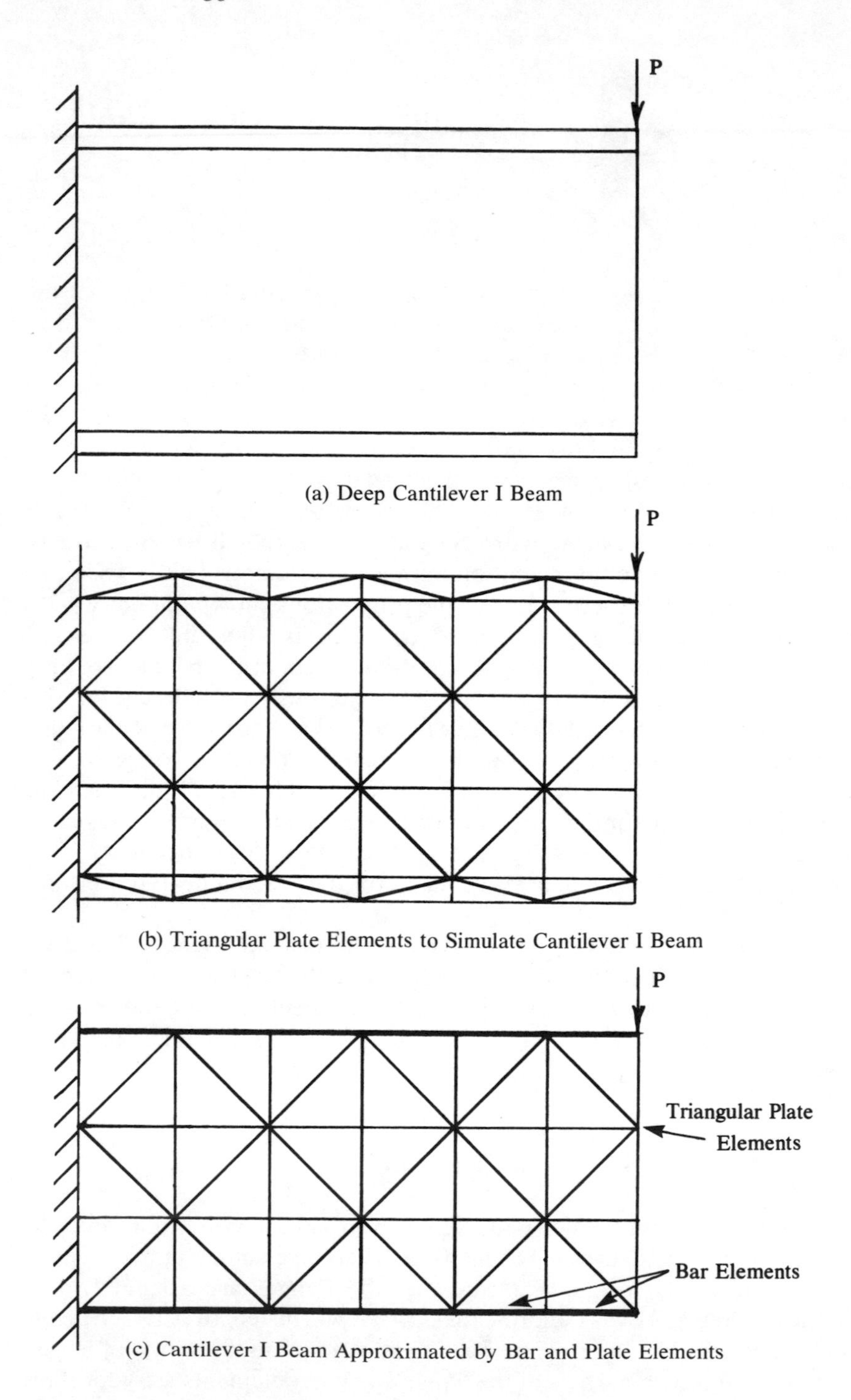

(a) Deep Cantilever I Beam

(b) Triangular Plate Elements to Simulate Cantilever I Beam

(c) Cantilever I Beam Approximated by Bar and Plate Elements

Figure 4-1. Cantilevered I Beam

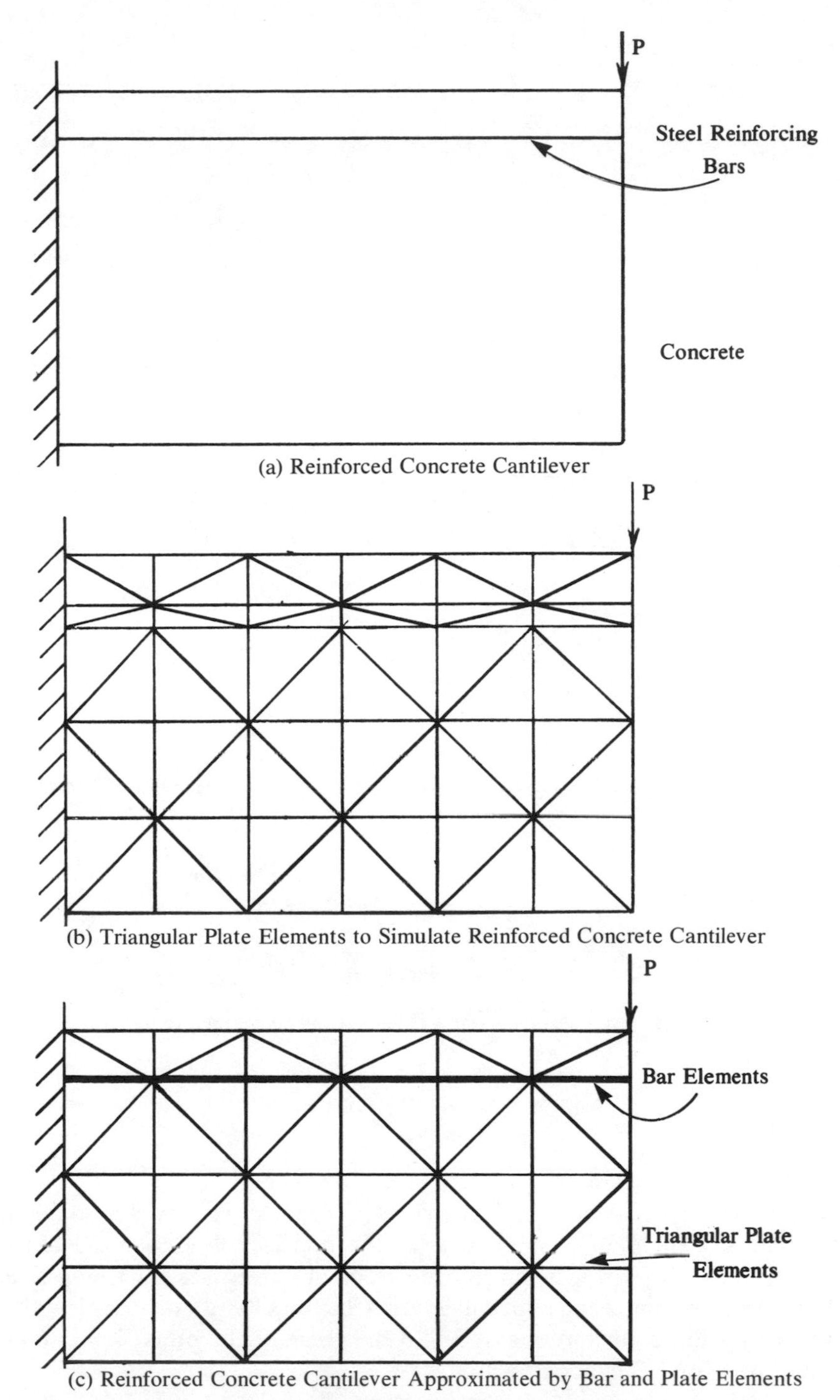

Figure 4-2. Reinforced Concrete Cantilever

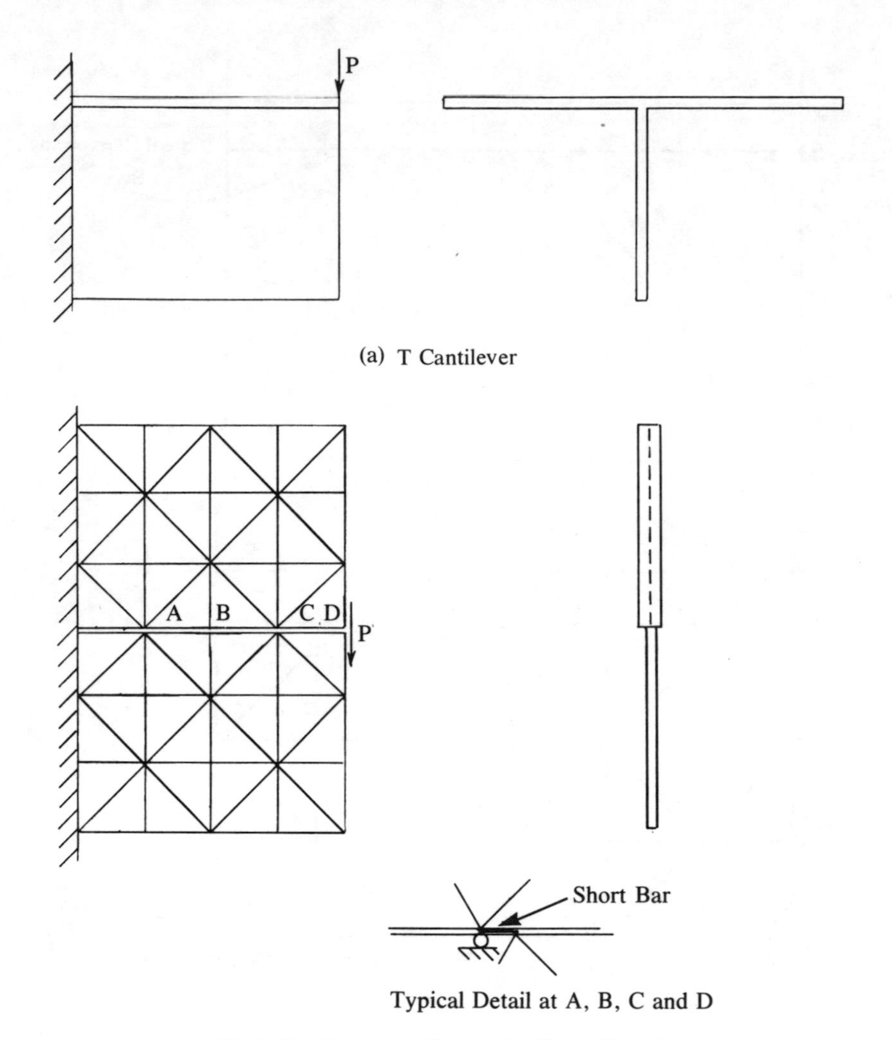

(a) T Cantilever

Typical Detail at A, B, C and D

(b) T Cantilever as a Composite Planar Structure

Figure 4-3. Shear-Lag Problem

in regions of rapid change in moment. Variation in stress through the thickness of the flange can be allowed for in interpreting the bar stresses.

To form the stiffness matrix, the element stiffnesses for triangles and bars are determined and added to S as in Chapters 1 and 2 or 3, each without regard for the contributions to S that members of the other family make. Thus, the total S is accumulated.

```
1  READ TITLE
   READ AND PRINT CASE TITLE
   READ AND PRINT Y.M. AND P.R. FOR PLATE MATERIAL
   SET CONSTANT VALUES IN D, B AND BDB                       DB03B
   CALCULATE DB
   READ COORDINATES OF NODES                                 RC06B
   READ PLATE ELEMENT DATA
   READ AND PRINT Y.M. OF BAR MATERIAL
   READ BAR ELEMENT DATA                                     CN10B
   ZERO TRIG AND S
   REWIND 1
   DO FOR ALL PLATE ELEMENTS
     DETERMINE A^-1 X AND Y FOR ORIGIN AT CENTROID           AI02B
     WRITE A^-1 ON 1                                         (IN12B)
     DETERMINE ELEMENT STIFFNESS, E                          ES06B
     ADD E TO S                                              AS06B
   DO FOR ALL BAR ELEMENTS
     CALCULATE STIFFNESS AND ADD TO S                        ES10B
   REWIND 4
   WRITE S ON 4
2  READ KNOWN FORCES                                         KF01B
   READ KNOWN DISPLACEMENTS                                  KU01B
   READ GUIDED NODE DATA                    }
   SET IFX COMPONENTS                       }                GN02B
   ALTER, FOR ROTATED AXES; F->F' AND S->S' }
   SOLVE FOR UNKNOWN COMPONENTS IN U'                        GE02B
   REWIND 4
   READ S FROM 4
   RESTORE U'->U                                             RU01B
   SOLVE FOR ALL FORCE COMPONENTS, PRINT F AND U             FU10B
   REWIND 1
   DO FOR ALL PLATE ELEMENTS
     READ A^-1 FROM 1
     CALCULATE AND PRINT STRESSES AT CENTROID                ST06B
   CALCULATE AND PRINT STRESSES IN BARS                      ST10B
   READ NEXT

                NEXT:1
         <        =         >
   CALL EXIT   GO TO 1    GO TO 2
```

Figure 4-4. SP44B Flow Chart

Stresses in a Composite Structure

When S has been formed, the loads, known displacement, and guided nodes are treated as previously described, and all displacements obtained. For elements in each family, the **δ** displacements can be extracted from **U** and, by the applicable stress-displacement formulas, the stresses found.

Other Applications

The power of the composite structure can be readily seen from the illustrations that have been used. Application to stiffened plate girders and castellated beams is obvious, and there are other not-so-obvious uses.

In the illustrations, the two types of elements were intimately connected; but they could have been quite separate, as in the case of a structure that consists of a plate connected to a pin-jointed truss. No doubt there are innumerable uses for composites; one more illustration will be given.

Suppose that the bending stresses in the T cantilever of Figure 4-3a are required. The simple beam formula will be unreliable because of shear-lag in the flange. If the flange is thin enough so that it can be considered to offer no significant resistance to small vertical deflections at the junction with the web, then the flange applies forces to the web that are horizontal and axial. These forces could equally well be transmitted between neighboring points by small bar elements. In that case, the effect of the flange is not altered if the plane of the flange is rotated from horizontal about the flange-web joining line. We can, thus, change the problem from a T to two coplanar plates joined by a series of short bars. The same effect can be obtained and a simpler problem encountered by folding the two sides of the flange and providing roller supports as shown in Figure 4-3b. In this way, a three-dimensional problem is reduced to a rather straight-forward planar problem that can be solved as a composite structure.

A Program to Solve Reinforced Plates by Bar Elements and Linear Strain Triangles

This program, SP44B, will not be described in detail, as it is in essence a merger of PT10B with SP33B. Reference should be made to Figure 4-4, SP44B Flow Chart; Figure 4-5, SP44B Fortran Statements; Figure 4-6, Instructions for SP44B Data Deck Preparations; and the computer output given in Figure 4-9.

Example

A 5×3 I beam weighing 10 lb/ft supporting a load of 5000 lb when cantilevered from a rigid wall, as shown in Figure 4-7, has been selected to show the usefulness of combining plate and bar elements. In order to indicate the accuracy of this method the finite element values of deflection and stress are compared with the theoretical values.

```
C          *****     MAIN SP44B     *****          JUNE 20,1972
C   STRESSES IN A PLATE USING LINEAR STRAIN TRIANGLES      W H BOWES
C    (REINFORCING BARS       GUIDED NODES)
      DIMENSION S(46,260),F(260),U(260),IFX(260),XY(2,130),TRIG(260)
      DIMENSION NCON(6,90),T(90)
      DIMENSION NCB(2,40),ARB(40)
      DIMENSION E(12,12),B(3,12),AI(12,12),X(3),Y(3),NAME(20)
      DIMENSION DB(3,12),D(3,3),BDB(12,12)
      LBAND=46
    1 WRITE(3,102)
      READ(2,100)NAME
      WRITE(3,101)NAME
      READ(2,103)YM,GNU
      WRITE(3,104)YM,GNU
      CALL DB03B(YM,GNU,D,B,BDB)
      DO 8 I=1,3
      DO 8 J=1,12
      DBIJ=0.
      DO 7 IJ=1,3
    7 DBIJ=DBIJ+D(I,IJ)*B(IJ,J)
    8 DB(I,J)=DBIJ
      CALL RC06B(2,XY,NDF)
      CALL  CN06B(LBAND,6,2,NCON,T,NE,NBAND)
      READ(2,103)YMB
      WRITE(3,107)YMB
      CALL CN10B(NCB,ARB,NB,NBAND)
      DO 9 J=1,NDF
      TRIG(J)=0.
      DO 9 I=1,NBAND
    9 S(I,J)=0.
      REWIND 1
      DO 15 IE=1,NE
      CALL AI02B(IE,NCON,XY,T,AI,X,Y)
      WRITE(1) AI
      V=T(IE)*.5*(X(1)*(Y(2)-Y(3))+X(2)*(Y(3)-Y(1))+X(3)*(Y(1)-Y(2)))
      CALL ES06B(V,AI,BDB,X,Y,E)
      CALL AS06B(IE,NCON,E,S,2,6,12,LBAND)
   15 CONTINUE
      END FILE 1
      IF(NB)21,21,19
   19 DO 20 IB=1,NB
   20 CALL ES10B(IB,NCB,XY,ARB,YMB,S,LBAND)
   21 REWIND 4
      DO 26 I=1,NBAND
   26 WRITE(4)(S(I,J),J=1,NDF)
      END FILE 4
    2 CALL KF01B(F,NDF)
      CALL KU01B(U,IFX,NDF)
      CALL GN02B(S,J,F,IFX,NBAND,LBAND,TRIG)
      CALL GE02B(F,J,S,IFX,NBAND,NDF,LBAND)
      REWIND 4
      DO 30 I=1,NBAND
   30 READ(4)(S(I,J),J=1,NDF)
      CALL RU01B(U,TRIG,NDF)
      CALL FU10B(S,U,NBAND,NDF,2,LBAND)
      REWIND 1
      WRITE(3,106)
      DO 45 IE=1,NE
      READ(1) AI
   45 CALL ST06B(DB,AI,U,NCON,IE)
      CALL ST10B(NB,NCB,XY,U,YMB)
      READ(2,108)NEXT
      IF(NEXT-1)47,1,2
   47 CALL EXIT
  100 FORMAT(20A4)
  101 FORMAT('0CASE TITLE  --- ',20A4)
  102 FORMAT('1              MAIN SP44B   JUNE 20 1972',/,' STRESSES IN IN-P
     1LANE LOADED PLATE ( LST   REINFORCING BARS     GUIDED NODES )')
  103 FORMAT(2F10.5)
  104 FORMAT('0YOUNGS MODULUS=',E10.3,'  POISSONS RATIO=',F6.3)
  106 FORMAT('0ELEM.NO.      SXX         SYY         SXY        THETA      PS1
     1      PS2')
  107 FORMAT('0YOUNGS MODULUS OF BARS =',E10.3)
  108 FORMAT(I5)
      END
```

Figure 4-5. SP44B Fortran Statements

Data Deck:

A card containing the Case Title.

A card containing the elastic constants of the plate;

Young's Modulus, Poisson's Ratio. Format: 2F10.5

A card containing the number of nodes in the system. Format: I5

One card for each node that is located at the corner of a triangle, containing;

Node number, x coordinate, y coordinate. Format: I5, 2F10.5

A blank card to indicate end of node data.

One card for each plate element containing;

Element number, Thickness (= 1 by default), Node numbers on periphery of element, in sequence around the periphery starting at a corner node. Format: I5, F10.5, 6I5

A blank card to indicate end of triangular element data.

A card containing Young's Modulus of the bar elements.

One card for each bar element, containing;

Element number, Cross-sectional area, Node number at each end of element. Format: I5, F10.5, 2I5

A blank card to indicate the end of the bar element data.

One card for each known, nonzero load component, containing;

Component number (twice node number minus one for an x component, or twice node number for a y component.) Magnitude of force, Format: I5, F10.5

A blank card to indicate end of load data.

One card for each known displacement component, containing;

Component number, Displacement. Format: I5, F10.5

A blank card to indicate end of displacement data.

One card for each guided node, containing;

Node number, Angle (alpha) in degrees defining the direction of the guiding plane, Normal displacement of guiding plane, Load tangential to guiding plane. Format: I5, 3F10.5

Note: Consider a set of axes, x′ and y′, rotated so that x′ is normal to the guiding plane. The angle of the x′ axis when measured counterclockwise from the x axis gives alpha. Normal displacement is positive if the plane moves in the +x′ direction. The load is positive if it is in the +y′ direction.

A blank card to indicate end of guided node data.

A card containing NEXT. Format: I5

NEXT = 0; End of job.

NEXT = 1; Execute a new case. Follow by data cards prepared according to all above instructions.

NEXT = 2; Repeat the case just completed but with a new set of known loads and displacements. Follow by data cards described above starting with force data cards.

Figure 4-6. Instruction for SP44B Data Deck Preparation

Problem

For the cantilevered I beam shown in Figure 4-7 determine the deflection and stress throughout the beam by the Finite Element Method and compare the results with the theoretical solution.

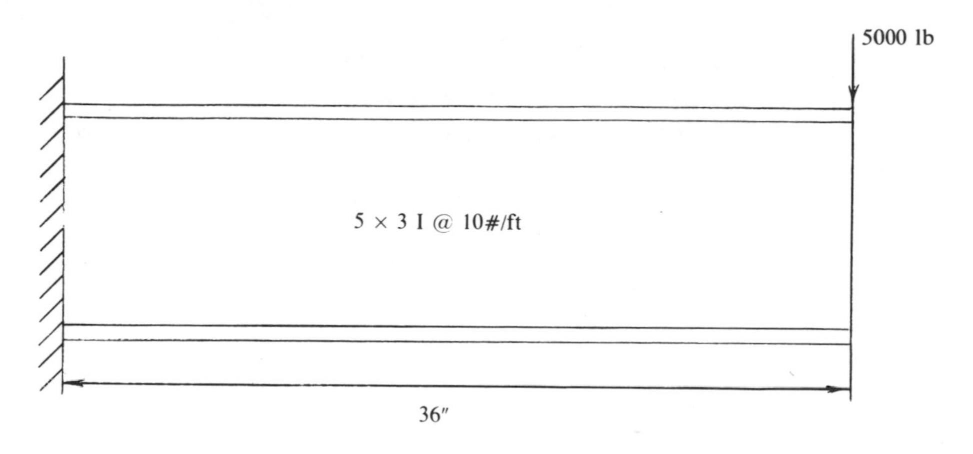

Figure 4-7. Cantilevered I Beam

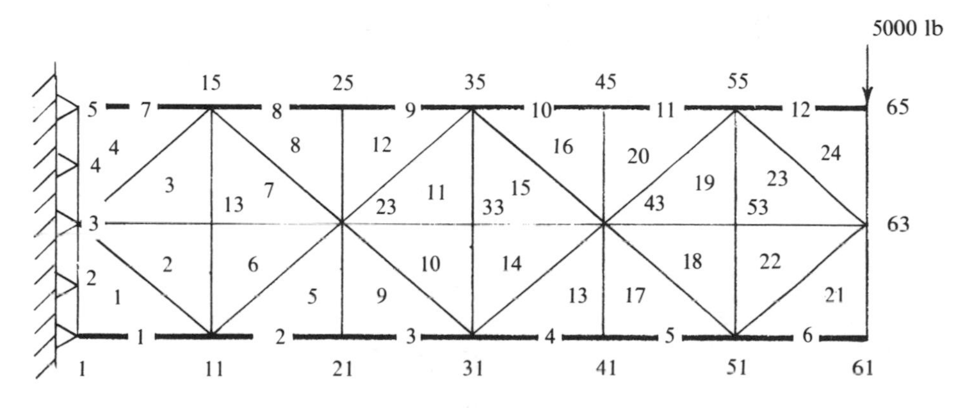

Figure 4-8. Element Subdivision for Cantilevered I Beam

Solution. The solution to this problem may be obtained using the Finite Element Method in a number of ways. The flanges of the I beam could be considered as triangular elements with a thickness equal to the width of the flange. A better approach, which is used here, is to treat the beam as a composite consisting of triangular elements for the web and bar elements for the flange. Such a subdivision of the beam is shown in Figure 4-8. The bar elements are numbered from 1 to 12 and the elements from 1 to 24. In numbering the nodal points, only those nodes for which information is read into the computer are numbered in Figure 4-8; that is, the nodes located at the mid point of the sides of the elements are omitted since the computer calculates their location automatically.

MAIN SP44B JUNE 20 1972
STRESSES IN IN-PLANE LOADED PLATE (LST REINFORCING BARS GUIDED NODES)

CASE TITLE --- STRESSES IN A CANTILEVERED I-BEAM

YOUNGS MODULUS= .300E+08 POISSONS RATIO= .300

NODE NO.	X-COORD	Y-COORD
1	.0000E+00	-.2337E+01
3	.0000E+00	.0000E+00
5	.0000E+00	.2337E+01
11	.6000E+01	-.2337E+01
13	.6000E+01	.0000E+00
15	.6000E+01	.2337E+01
21	.1200E+02	-.2337E+01
23	.1200E+02	.0000E+00
25	.1200E+02	.2337E+01
31	.1800E+02	-.2337E+01
33	.1800E+02	.0000E+00
35	.1800E+02	.2337E+01
41	.2400E+02	-.2337E+01
43	.2400E+02	.0000E+00
45	.2400E+02	.2337E+01
51	.3000E+02	-.2337E+01
53	.3000E+02	.0000E+00
55	.3000E+02	.2337E+01
61	.3600E+02	-.2337E+01
63	.3600E+02	.0000E+00
65	.3600E+02	.2337E+01

SYSTEM HAS 130 DEGREES OF FREEDOM

ELEM.NO.	THICKNESS	CONNECTING NODE NUMBERS					
1	.21	1	6	11	7	3	2
2	.21	11	12	13	8	3	7
3	.21	3	8	13	14	15	9
4	.21	15	10	5	4	3	9
5	.21	11	16	21	22	23	17
6	.21	23	18	13	12	11	17
7	.21	13	18	23	19	15	14
8	.21	23	24	25	20	15	19
9	.21	21	26	31	27	23	22
10	.21	31	32	33	28	23	27
11	.21	23	28	33	34	35	29
12	.21	23	29	35	30	25	24
13	.21	31	36	41	42	43	37
14	.21	31	37	43	38	33	32
15	.21	33	38	43	39	35	34
16	.21	43	44	45	40	35	39
17	.21	41	46	51	47	43	42
18	.21	51	52	53	48	43	47
19	.21	43	48	53	54	55	49
20	.21	43	49	55	50	45	44
21	.21	51	56	61	62	63	57
22	.21	51	57	63	58	53	52
23	.21	53	58	63	59	55	54
24	.21	63	64	65	60	55	59

BAND WIDTH = 26

YOUNGS MODULUS OF BARS = .300E+08

BAR NUMBER	X-SECT AREA	CONNECTS NODES NO.	
1	.98	1	11
2	.98	11	21
3	.98	21	31
4	.98	31	41

Figure 4-9. Output Processed by SP44B

```
          5               .98             41   51
          6               .98             51   61
          7               .98              5   15
          8               .98             15   25
          9               .98             25   35
         10               .98             35   45
         11               .98             45   55
         12               .98             55   65

 BAND WIDTH =    26

   KNOWN NON-ZERO LOADS
 COMPONENT NUMBER    LOAD
       130          -.5000E+04

       KNOWN DISPLACEMENTS
 COMPONENT NUMBER    DISPLACEMENT
             2          .0000E+00
             3          .0000E+00
             4          .0000E+00
             5          .0000E+00
             6          .0000E+00
             7          .0000E+00
             8          .0000E+00
             9          .0000E+00
            10          .0000E+00

NODE NO.   ALPHA(DEG)   KNOWN U     TANG.FORCE

NODE NO.         FORCE AND DISPLACEMENT COMPONENTS
      1       .3569E+05  .3668E+04 -.3817E-14 -.3919E-15
      2       .5649E+04 -.1079E+04 -.4751E-15  .4324E-16
      3      -.6563E-03 -.1774E+03  .7360E-22  .9478E-17
      4      -.5649E+04 -.1079E+04  .4751E-15  .4324E-16
      5      -.3569E+05  .3668E+04  .3817E-14 -.3919E-15
      6       .1819E-11  .4718E-10 -.3706E-02 -.4020E-02
      7       .2355E-10 -.1291E-10 -.1749E-02 -.3808E-02
      8      -.5334E-11  .2233E-10 -.9084E-10 -.3743E-02
      9      -.2456E-10 -.1424E-10  .1749E-02 -.3808E-02
     10       .2728E-11  .4330E-10  .3706E-02 -.4020E-02
     11       .8731E-10  .4815E-09 -.6151E-02 -.1185E-01
     12       .1016E-09  .4745E-09 -.3102E-02 -.1160E-01
     13       .3070E-11  .5248E-09  .2762E-09 -.1151E-01
     14      -.8297E-10  .4360E-09  .3102E-02 -.1160E-01
     15      -.1393E-09  .4802E-09  .6151E-02 -.1185E-01
     16       .3563E-10  .7138E-09 -.9420E-02 -.2335E-01
     17       .7276E-10  .8913E-09 -.4612E-02 -.2307E-01
     18      -.1654E-12  .7007E-09 -.1256E-08 -.2296E-01
     19      -.9955E-10  .7894E-09  .4612E-02 -.2307E-01
     20      -.2728E-11  .5866E-09  .9420E-02 -.2335E-01
     21       .3156E-09  .1822E-08 -.1122E-01 -.3800E-01
     22       .5694E-10  .1696E-08 -.5671E-02 -.3781E-01
     23      -.7822E-10  .1617E-08 -.7060E-08 -.3774E-01
     24      -.1669E-09  .1072E-08  .5671E-02 -.3781E-01
     25      -.3311E-09  .1280E-08  .1122E-01 -.3800E-01
     26      -.2910E-10  .1358E-08 -.1392E-01 -.5554E-01
     27       .1003E-09  .2205E-08 -.6863E-02 -.5544E-01
     28       .1675E-10  .1929E-08 -.1403E-06 -.5541E-01
     29      -.1601E-09  .1406E-08  .6862E-02 -.5544E-01
     30       .6542E-10  .1080E-08  .1392E-01 -.5554E-01
     31       .4657E-09  .4545E-08 -.1515E-01 -.7572E-01
     32       .1579E-09  .3036E-08 -.7608E-02 -.7557E-01
     33       .7047E-10  .5054E-08 -.2352E-06 -.7552E-01
     34      -.1785E-09  .3970E-08  .7608E-02 -.7557E-01
     35      -.4175E-09  .3775E-08  .1515E-01 -.7572E-01
     36      -.7406E-10  .2315E-08 -.1729E-01 -.9805E-01
     37       .1455E-10  .3234E-08 -.8551E-02 -.9787E-01
```

Figure 4-9 (continued)

38	.2757E-10	.1834E-08	-.1816E-05	-.9780E-01
39	-.3153E-09	.3026E-08	.8548E-02	-.9786E-01
40	.5821E-10	.2171E-08	.1729E-01	-.9805E-01
41	.5239E-09	.6287E-08	-.1796E-01	-.1221E+00
42	.1394E-09	.5275E-08	-.9049E-02	-.1220E+00
43	-.1310E-09	.7501E-08	-.6723E-05	-.1220E+00
44	-.1391E-09	.1706E-08	.9039E-02	-.1220E+00
45	-.3929E-09	.4968E-08	.1797E-01	-.1221E+00
46	-.3056E-09	.5495E-08	-.1956E-01	-.1476E+00
47	-.6255E-11	.5582E-08	-.9693E-02	-.1476E+00
48	.1291E-10	.4082E-08	-.2463E-04	-.1476E+00
49	-.2183E-09	.4632E-08	.9654E-02	-.1475E+00
50	.3695E-09	.8237E-09	.1953E-01	-.1476E+00
51	.5821E-09	.8259E-08	-.1963E-01	-.1743E+00
52	.2202E-09	.7289E-08	-.9904E-02	-.1742E+00
53	.1274E-09	.1153E-07	-.7700E-04	-.1742E+00
54	-.9503E-10	.5191E-08	.9804E-02	-.1742E+00
55	-.4802E-09	.8104E-08	.1966E-01	-.1742E+00
56	-.3077E-09	.7756E-08	-.2071E-01	-.2016E+00
57	-.7276E-10	.8064E-08	-.1038E-01	-.2015E+00
58	.1350E-10	.5573E-08	-.1664E-03	-.2015E+00
59	-.9507E-10	.7195E-08	.1005E-01	-.2015E+00
60	.0000E+00	.4571E-08	.2054E-01	-.2016E+00
61	-.2910E-09	.1154E-08	-.2012E-01	-.2285E+00
62	-.2459E-09	.5317E-08	-.1026E-01	-.2288E+00
63	-.2994E-09	.9497E-08	-.2381E-03	-.2292E+00
64	.1641E-09	.2995E-08	.1005E-01	-.2296E+00
65	.3638E-09	-.5000E+04	.2033E-01	-.2301E+00

ELEM.NO.	SXX	SYY	SXY	THETA	PS1	PS2
1	-.229E+05	-.322E+04	-.447E+04	12.2	-.238E+05	-.225E+04
2	-.908E+04	-.194E+03	-.571E+04	26.1	-.119E+05	.260E+04
3	.908E+04	.194E+03	-.571E+04	-26.1	.119E+05	-.260E+04
4	.229E+05	.322E+04	-.447E+04	-12.2	.238E+05	.225E+04
5	-.151E+05	.150E+04	-.445E+04	14.1	-.162E+05	.262E+04
6	-.100E+05	.184E+03	-.574E+04	24.2	-.126E+05	.276E+04
7	.100E+05	-.184E+03	-.574E+04	-24.2	.126E+05	-.276E+04
8	.151E+05	-.150E+04	-.445E+04	-14.1	.162E+05	-.262E+04
9	-.149E+05	-.150E+04	-.446E+04	16.8	-.163E+05	-.158E+03
10	-.502E+04	-.159E+03	-.573E+04	33.5	-.882E+04	.364E+04
11	.502E+04	.157E+03	-.573E+04	-33.5	.881E+04	-.364E+04
12	.149E+05	.150E+04	-.446E+04	-16.8	.163E+05	.159E+03
13	-.756E+04	.153E+04	-.448E+04	22.3	-.940E+04	.337E+04
14	-.624E+04	.169E+03	-.574E+04	30.4	-.961E+04	.354E+04
15	.621E+04	-.148E+03	-.573E+04	-30.5	.958E+04	-.352E+04
16	.754E+04	-.147E+04	-.443E+04	-22.3	.935E+04	-.329E+04
17	-.751E+04	-.152E+04	-.453E+04	28.2	-.994E+04	.917E+03
18	-.152E+04	.123E+02	-.594E+04	41.3	-.675E+04	.524E+04
19	.987E+03	.347E+03	-.553E+04	-43.3	.620E+04	-.487E+04
20	.737E+04	.149E+04	-.438E+04	-28.1	.971E+04	-.843E+03
21	-.492E+03	-.157E+04	-.403E+04	-41.2	.303E+04	-.510E+04
22	-.348E+04	-.206E+03	-.585E+04	37.2	-.792E+04	.423E+04
23	.123E+04	-.553E+03	-.559E+04	-40.5	.600E+04	-.533E+04
24	-.730E+03	-.602E+04	-.491E+04	-30.8	.220E+04	-.895E+04

BAR NO.	STRESS PSI
1	-.3076E+05
2	-.2534E+05
3	-.1963E+05
4	-.1409E+05
5	-.8305E+04
6	-.2472E+04
7	.3076E+05
8	.2534E+05
9	.1963E+05
10	.1411E+05
11	.8475E+04
12	.3345E+04

EXIT

Figure 4-9 (continued)

FINITE ELEMENT RESULTS

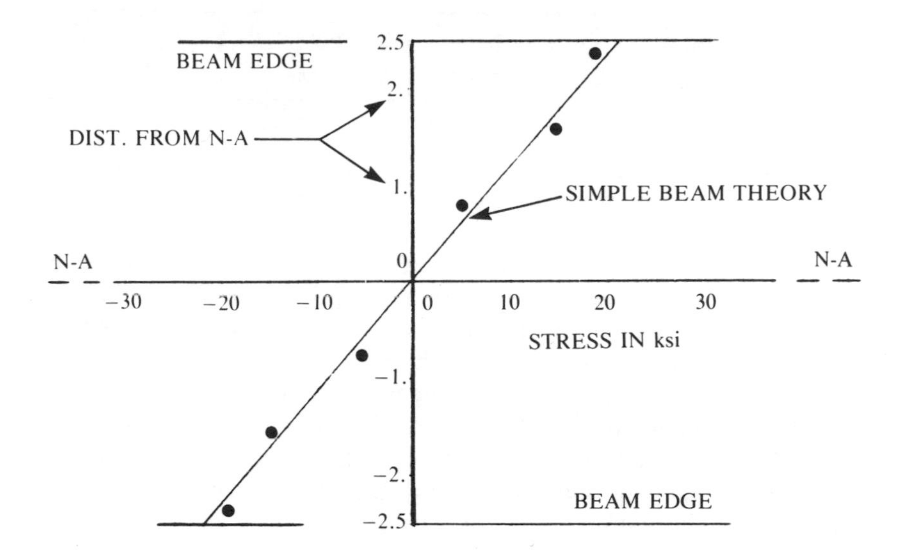

Figure 4-10. Flexural Stress in I Beam

The boundary conditions for this cantilevered beam are represented by pin connections at nodes 1 to 5 inclusive, thus, constraining the Finite Element Method model in the same manner as the original beam.

The loading for this example, 5000 lb applied at the free end of the beam, can be applied to the Finite Element Method model as a vertical nodal point load at node 65 as shown in Figure 4-8.

Results

The computer output for this example is shown in Figure 4-9, and is interpreted in Figure 4-10 and Table 4-1.

In order to show the accuracy obtained using linear strain triangles, combined with bar elements in this example, the theoretical values of the flexural stress and the finite element values are compared in Figure 4-10. Close agreement is obtained, with only four triangular elements and two bar elements used to represent the full depth of the beam, between the theoretical values and the Finite Element Method computer results. The percentage error of the stress at the center of gravity of the flange, namely 3.2 percent, could be further reduced by increasing the number of elements across the depth of the beam.

Table 4-1
Deflections at Free End of I Beam

	Deflections in Inches		
	Bending	*Shear*	*Total*
Theoretical	−.2142	−.0171	−.2312
Finite Element Method	—	—	−.2301

A further indication of the accuracy is shown in Table 4-1 where the deflections at the free end of the beam as found by the Finite Element Method and theoretically are recorded. The method gives a deflection which is, in fact, more accurate than the theoretical value for the free end of the beam when the deflection due to shear is not considered. When the shear deflection is combined with the deflection due to bending, the Finite Element Method value is within 1/2 percent of the theoretical value.

The results of this example, even though the number of elements used is small, indicates the nature of the accuracy obtainable using the Finite Element Method.

5 Axi-Symmetric Shells

The general shell problem encountered in the analysis of ship's plates and in the skin of aircraft wings is a three-dimensional problem and, hence, is very difficult to solve. It would be unwise to undertake such a solution before solving simpler introductory shell problems. To introduce shells, we will now consider the case of an axi-symmetric shell of uniform thickness supporting an axi-symmetric load. This is as far as our treatment of shells will go. Extension to the case of the shell of varying thickness is quite simple, but this still falls far short of the perfectly general case.

Within the realm of the axi-symmetric shell, there are many problems of interest to engineers. The mechanical engineer is interested in the stresses in and near the heads of pressure vessels such as that of Figure 5-1a. The bending stresses in the doubly curved portion of the heads and at the welds, which are not amenable to analytical solutions, are of particular importance. Civil engineers need to evaluate stresses in shells such as the roof shown in Figure 5-1b. Analytical solutions are available for only a very limited range of shapes and edge conditions [13].

In this analysis, the axis of symmetry will be coincident with the x axis and the shell will be divided into a number of conical elements. More refined curved elements have been used [9] to obtain excellent results; but only the conical element, which gives good results when a large number of elements are used, will be treated here.

The Conical Element

Consider that the stresses are required in a shell as shown in Figure 5-2a and that the load, as well as the shell, has complete axial symmetry. The x axis is made to coincide with the axis of symmetry while the y axis is placed at a convenient location. Let the shell be approximated by a number of conical elements arranged so that the elements are small in regions of large curvature and in regions of large bending moment. A single typical element such as that shown in Figure 5-2b will be considered and its stiffness calculated.

The profile of the element in its unloaded and loaded positions is shown in Figure 5-3. The displacements are given by $\boldsymbol{\delta}'$ at the ends and $\begin{Bmatrix} u' \\ v' \\ \beta' \end{Bmatrix}$ at

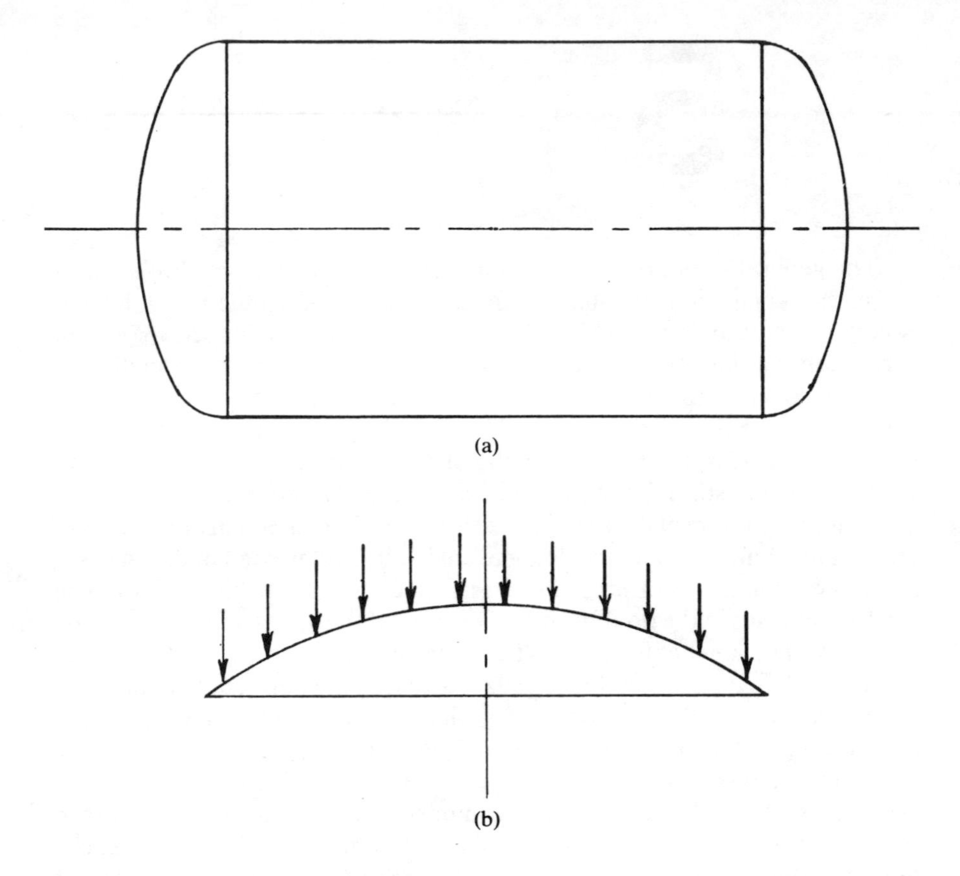

Figure 5-1. Shell Problems

a typical position located by x'. It will be noted that rotation is included for the first time. This is necessary because compatibility of axial displacements at the nodes is not sufficient to give acceptable accuracy in the solution and compatibility of rotation must be assured. Henceforth, the word displacement will be used to indicate components of rotation as well as translation.

Displacement Functions

Again, polynomial displacement functions will be assumed; the polynomials being of the highest order that will permit evaluation of the constant coefficients. Since the displacement u' is known at two points, nodes 1 and 2, the highest-order expression which can be assumed for u' is

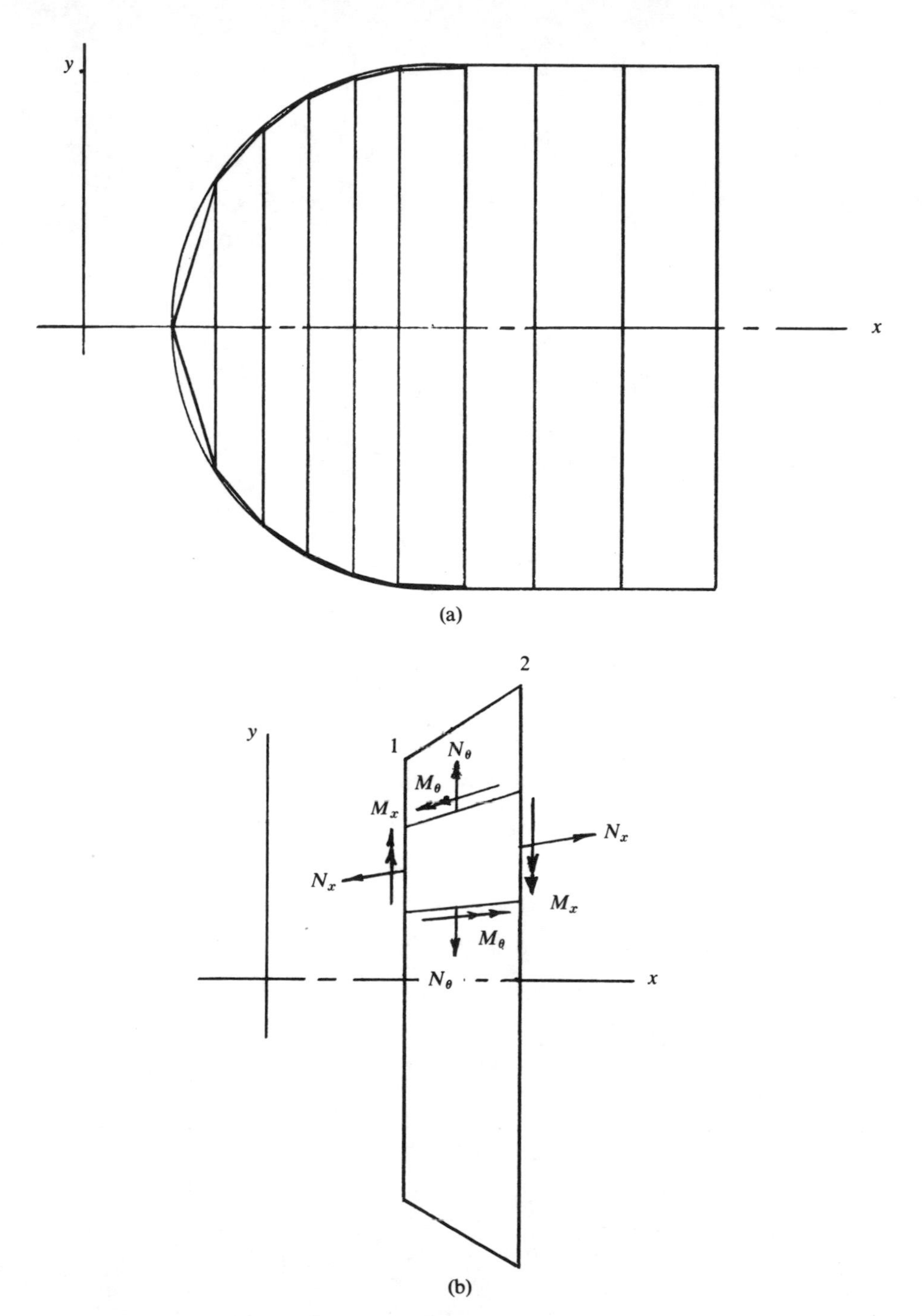

Figure 5-2. Shell Element

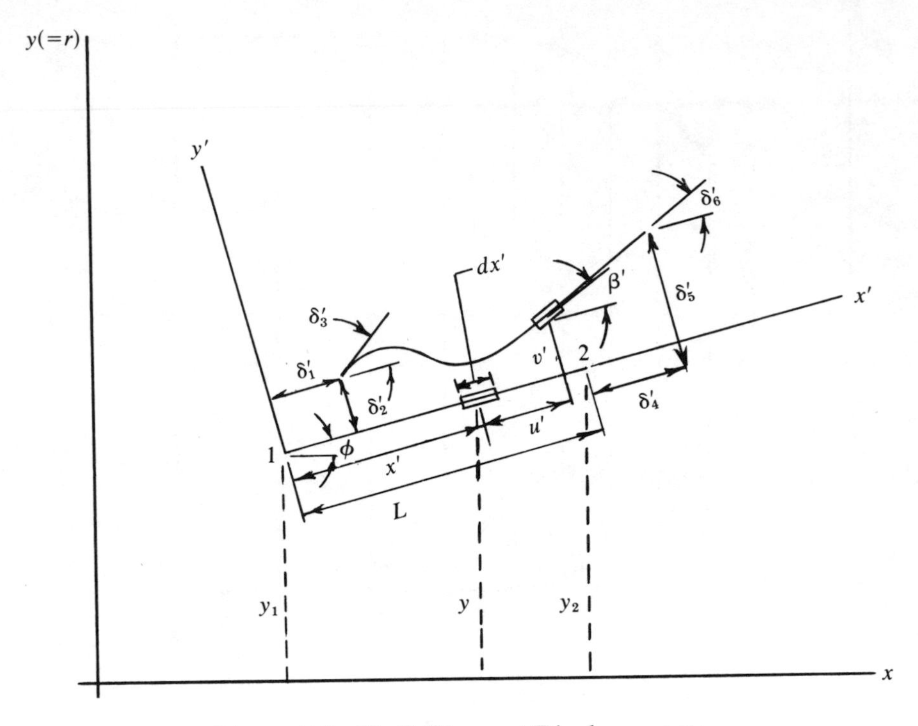

Figure 5-3. Shell Element Displacements

$$u' = \alpha_1 + \alpha_2 x'$$

The v' displacement with its derivative has four known values; hence, it may be assumed that

$$v' = \alpha_3 + \alpha_4 x' + \alpha_5 x'^2 + \alpha_6 x'^3$$

From which it follows that

$$\beta' = \alpha_4 + 2\alpha_5 x' + 3\alpha_6 x'^2$$

These functions can be combined into

$$\begin{Bmatrix} u' \\ v' \\ \beta' \end{Bmatrix} = \begin{bmatrix} 1 & x'^1 & 0 & 0 & 0 & 0 \\ 0 & 0 & 1 & x'^1 & x'^2 & x'^3 \\ 0 & 0 & 0 & 1 & 2x'^1 & 3x'^2 \end{bmatrix} \begin{Bmatrix} \alpha_1 \\ \alpha_2 \\ \alpha_3 \\ \alpha_4 \\ \alpha_5 \\ \alpha_6 \end{Bmatrix} \tag{5.1}$$

Substituting the coordinates of the ends of the element into (5.1) gives

$$\begin{Bmatrix} \delta_1' \\ \delta_2' \\ \delta_3' \\ \delta_4' \\ \delta_5' \\ \delta_6' \end{Bmatrix} = \begin{bmatrix} 1 & 0 & 0 & 0 & 0 & 0 \\ 0 & 0 & 1 & 0 & 0 & 0 \\ 0 & 0 & 0 & 1 & 0 & 0 \\ 1 & L & 0 & 0 & 0 & 0 \\ 0 & 0 & 1 & L & L^2 & L^3 \\ 0 & 0 & 0 & 1 & 2L & 3L^2 \end{bmatrix} \begin{Bmatrix} \alpha_1 \\ \alpha_2 \\ \alpha_3 \\ \alpha_4 \\ \alpha_5 \\ \alpha_6 \end{Bmatrix}$$

or

$$\boldsymbol{\delta}' = \mathbf{C}\boldsymbol{\alpha} \tag{5.2}$$

where

$$\mathbf{C} = \begin{bmatrix} 1 & 0 & 0 & 0 & 0 & 0 \\ 0 & 0 & 1 & 0 & 0 & 0 \\ 0 & 0 & 0 & 1 & 0 & 0 \\ 1 & L & 0 & 0 & 0 & 0 \\ 0 & 0 & 1 & L & L^2 & L^3 \\ 0 & 0 & 0 & 1 & 2L & 3L^2 \end{bmatrix} \tag{5.3}$$

From (5.2), the unknown coefficients can be determined by $\boldsymbol{\alpha} = \mathbf{C}^{-1}\boldsymbol{\delta}'$ which, when substituted into (5.1), gives

$$\begin{Bmatrix} u' \\ v' \\ \beta' \end{Bmatrix} = \begin{bmatrix} 1 & x' & 0 & 0 & 0 & 0 \\ 0 & 0 & 1 & x' & x'^2 & x'^3 \\ 0 & 0 & 0 & 1 & 2x' & 3x'^2 \end{bmatrix} \mathbf{C}^{-1}\boldsymbol{\delta}' \tag{5.4}$$

Stress-Strain Relations

Stress varies from point to point along the shell profile and also throughout the thickness of the shell. It is, thus, in reality an unknown function of two variables. To eliminate one of the variables, we will assume that bending stress varies linearly from a neutral surface which is equidistant from the inner and outer shell surfaces. This is a good approximation provided that the thickness is small in comparison with the radius of curvature [13]. Instead of dealing with this varying bending stress in the determination of strain energy—which would require integrating through the shell thickness—we will deal with bending moment as a "stress." The assumed linear variation of bending stress through the shell leads to the well-known relationship between bending moment per unit length, M, and bending stress at the surface.

$$\sigma = 6M/t^2 \tag{5.5a}$$

As well as bending, there are normal loads acting on any cross-section. Stress due to the normal load may be taken as constant throughout the thickness and can therefore be treated more readily than the bending stress. However, to be consistent with the treatment of bending, we will consider this membrane "stress" to be force per unit length, N, which is related to stress by

$$\sigma = N/t \tag{5.5b}$$

There are normal and bending stresses in the x' direction and in the circumferential direction; hence, the four components of "stress" are

$$\boldsymbol{\sigma}' = \begin{Bmatrix} N_{x'} \\ N_\theta \\ M_{x'x'} \\ M_{\theta\theta} \end{Bmatrix}$$

This vector will be treated as the stress vector to be evaluated. When the values of the stress vector have been found, the true stresses readily follow from (5.5).

The components of strain must now be selected. Again several alternatives are open, but we would prefer the strain energy to be evaluated, as previously, by

$$\int \boldsymbol{\varepsilon}^T \boldsymbol{\sigma}\, dv$$

This will enable us to again use the general formulas developed in Chapter 2. The consistent strain components are:

$$\boldsymbol{\varepsilon} = \begin{Bmatrix} \varepsilon_{x'x'} \\ \varepsilon_{\theta\theta} \\ \chi_{x'x'} \\ \chi_{\theta\theta} \end{Bmatrix}$$

where χ is the change in curvature that occurs during loading.

From elementary plate theory we have

$$N_{x'} = \frac{E}{1-\nu^2} t\varepsilon_{x'x'} + \frac{E}{1-\nu^2} t\nu\varepsilon_{\theta\theta}$$

$$N_{\theta} = \frac{E}{1-\nu^2} t\nu\varepsilon_{x'x'} + \frac{E}{1-\nu^2} t\varepsilon_{\theta\theta}$$

$$M_{x'x'} = \frac{E}{1-\nu^2}\frac{t^3}{12}\chi_{x'x'} + \frac{E}{1-\nu^2}\frac{t^3}{12}\nu\chi_{\theta\theta}$$

$$M_{\theta\theta} = \frac{E}{1-\nu^2}\frac{t^3}{12}\nu\chi_{x'x'} + \frac{E}{1-\nu^2}\frac{t^3}{12}\chi_{\theta\theta}$$

which can be written:

$$\begin{Bmatrix} N_{x'} \\ N_{\theta} \\ M_{x'x'} \\ M_{\theta\theta} \end{Bmatrix} = \frac{Et}{1-\nu^2}\begin{bmatrix} 1 & \nu & 0 & 0 \\ \nu & 1 & 0 & 0 \\ 0 & 0 & t^2/12 & \nu t^2/12 \\ 0 & 0 & \nu t^2/12 & t^2/12 \end{bmatrix}\begin{Bmatrix} \varepsilon_{x'x'} \\ \varepsilon_{\theta\theta} \\ \chi_{x'x'} \\ \chi_{\theta\theta} \end{Bmatrix}$$

or

$$\boldsymbol{\sigma} = \mathbf{D}\boldsymbol{\varepsilon} \tag{5.6}$$

where

$$\mathbf{D} = \frac{Et}{1-\nu^2}\begin{bmatrix} 1 & \nu & 0 & 0 \\ \nu & 1 & 0 & 0 \\ 0 & 0 & t^2/12 & \nu t^2/12 \\ 0 & 0 & \nu t^2/12 & t^2/12 \end{bmatrix} \tag{5.7}$$

Strain-Displacement Relations

Strain in the x' direction, in terms of displacement, is simply

$$\varepsilon_{x'x'} = \frac{\partial u'}{\partial x'}$$

The next pair of strain components are also quite simple:

$$\varepsilon_{\theta\theta} = \frac{v}{y} = \frac{u' \sin \phi}{y} + \frac{v' \cos \phi}{y}$$

$$\chi_{x'x'} = \frac{\partial \beta'}{\partial x'}$$

but determination of the last component is more difficult.

Consider a portion of the element as shown in Figure 5-4. The initial position and shape of the segment is shown in solid lines. After straining, as well as the u' and v' displacements, there is rotation through an angle β' to the state shown by the broken lines. The radius of curvature changes from R_0 to:

$$R = R_0 + \Delta R$$

From the trigonometry of Figure 5-4, we have

$$\Delta R = R_0 \beta' \tan \phi$$

The change of curvature is given by

$$\begin{aligned} \chi_{\theta\theta} &= \frac{1}{R_0} - \frac{1}{R} \\ &= \frac{R - R_0}{RR_0} = \frac{\Delta R}{RR_0} \simeq \frac{\Delta R}{R_0^2} \\ &= \frac{R_0 \beta' \tan \phi}{R_0^2} = \frac{\beta' \tan \phi}{R_0} = \frac{\beta' \tan \phi}{y/\cos \phi} \\ &= \frac{\beta' \sin \phi}{y} \end{aligned}$$

Summarizing:

$$\boldsymbol{\varepsilon} = \begin{Bmatrix} \varepsilon_{x'x'} \\ \varepsilon_{\theta\theta} \\ \chi_{x'x'} \\ \chi_{\theta\theta} \end{Bmatrix} = \begin{Bmatrix} \dfrac{\partial u'}{\partial x'} \\ \dfrac{\sin \phi u'}{y} + \dfrac{\cos \phi v'}{y} \\ \dfrac{\partial \beta'}{\partial x'} \\ \dfrac{\sin \phi}{y} \beta' \end{Bmatrix}$$

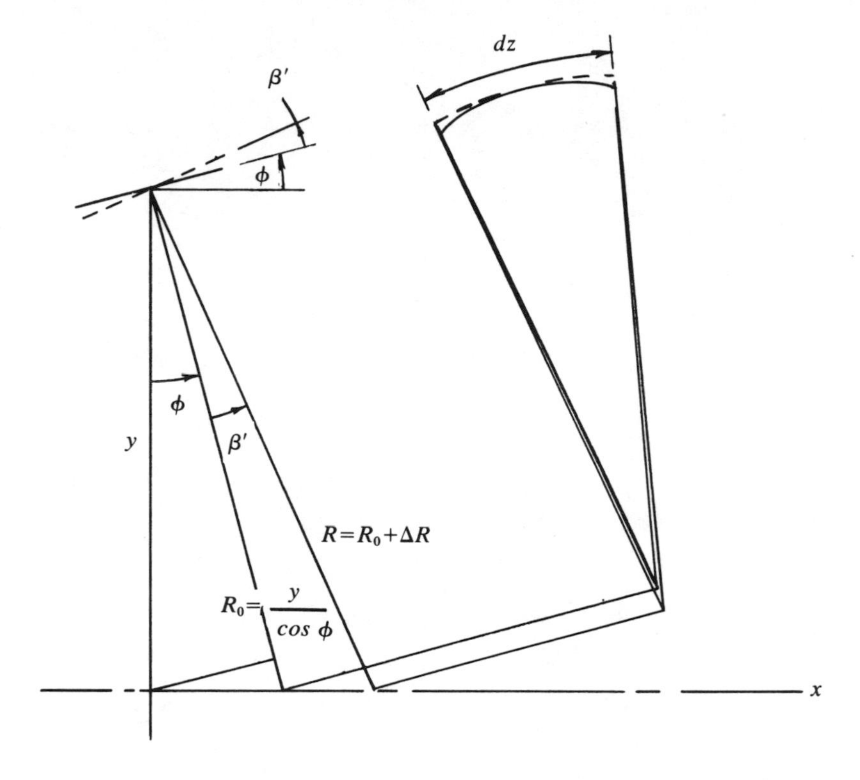

Figure 5-4. Segment of Conical Element

$$\boldsymbol{\varepsilon} = \begin{bmatrix} \dfrac{\partial}{\partial x'} & 0 & 0 \\ \dfrac{\sin\phi}{y} & \dfrac{\cos\phi}{y} & 0 \\ 0 & 0 & \dfrac{\partial}{\partial x'} \\ 0 & 0 & \dfrac{\sin\phi}{y} \end{bmatrix} \begin{Bmatrix} u' \\ v' \\ \beta' \end{Bmatrix}$$

Substituting from (5.4) gives

$$\boldsymbol{\varepsilon} = \begin{bmatrix} \dfrac{\partial}{\partial x'} & 0 & 0 \\ \dfrac{\sin\phi}{y} & \dfrac{\cos\phi}{y} & 0 \\ 0 & 0 & \dfrac{\partial}{\partial x'} \\ 0 & 0 & \dfrac{\sin\phi}{y} \end{bmatrix} \begin{bmatrix} 1 & x' & 0 & 0 & 0 & 0 \\ 0 & 0 & 1 & x' & x'^2 & x'^3 \\ 0 & 0 & 0 & 1 & 2x' & 3x'^2 \end{bmatrix} \mathbf{C}^{-1}\boldsymbol{\delta}'$$

$$= \begin{bmatrix} 0 & 1 & 0 & 0 & 0 & 0 \\ \dfrac{\sin\phi}{y} & \dfrac{x'\sin\phi}{y} & \dfrac{\cos\phi}{y} & \dfrac{x'\cos\phi}{y} & \dfrac{x'^2\cos\phi}{y} & \dfrac{x'^3\cos\phi}{y} \\ 0 & 0 & 0 & 0 & 2 & 6x' \\ 0 & 0 & 0 & \dfrac{\sin\phi}{y} & \dfrac{2x'\sin\phi}{y} & \dfrac{3x'^2\sin\phi}{y} \end{bmatrix} \mathbf{C}^{-1}\boldsymbol{\delta}'$$

$$= \mathbf{B}\mathbf{C}^{-1}\boldsymbol{\delta}' \tag{5.8}$$

where

$$\mathbf{B} = \begin{bmatrix} 0 & 1 & 0 & 0 & 0 & 0 \\ \dfrac{\sin\phi}{y} & \dfrac{x'\sin\phi}{y} & \dfrac{\cos\phi}{y} & \dfrac{x'\cos\phi}{y} & \dfrac{x'^2\cos\phi}{y} & \dfrac{x'^3\cos\phi}{y} \\ 0 & 0 & 0 & 0 & 2 & 6x' \\ 0 & 0 & 0 & \dfrac{\sin\phi}{y} & \dfrac{2x'\sin\phi}{y} & \dfrac{3x'^2\sin\phi}{y} \end{bmatrix} \tag{5.9}$$

Stiffness of a Conical Element

Following the same procedure as in Chapter 2, where the internal strain energy is equated to the work done by the external forces during a virtual displacement, will lead to

$$\mathbf{f}' = [\mathsf{C}^{-1}]^T \int \mathsf{B}^T \mathsf{D} \mathsf{B} \, d_{\text{vol}} \mathsf{C}^{-1} \boldsymbol{\delta}' \tag{5.10}$$

But force and displacement components in the x' and y' directions cannot be used in the assembled shell; hence, $\mathbf{f}'$ and $\boldsymbol{\delta}'$ must be transformed to $\mathbf{f}$ and $\boldsymbol{\delta}$.

Considering components, it is obvious that

$$\begin{Bmatrix} \delta_1' \\ \delta_2' \\ \delta_3' \\ \delta_4' \\ \delta_5' \\ \delta_6' \end{Bmatrix} = \begin{bmatrix} \cos\phi & \sin\phi & 0 & 0 & 0 & 0 \\ -\sin\phi & \cos\phi & 0 & 0 & 0 & 0 \\ 0 & 0 & 1 & 0 & 0 & 0 \\ 0 & 0 & 0 & \cos\phi & \sin\phi & 0 \\ 0 & 0 & 0 & -\sin\phi & \cos\phi & 0 \\ 0 & 0 & 0 & 0 & 0 & 1 \end{bmatrix} \begin{Bmatrix} \delta_1 \\ \delta_2 \\ \delta_3 \\ \delta_4 \\ \delta_5 \\ \delta_6 \end{Bmatrix}$$

or

$$\boldsymbol{\delta}' = \begin{bmatrix} \lambda & 0 \\ 0 & \lambda \end{bmatrix} \boldsymbol{\delta} \tag{5.11}$$

where

$$[\lambda] = \begin{bmatrix} \cos\phi & \sin\phi & 0 \\ -\sin\phi & \cos\phi & 0 \\ 0 & 0 & 1 \end{bmatrix} \tag{5.12}$$

Similarly, for force components,

$$\mathbf{f}' = \begin{bmatrix} \lambda & 0 \\ 0 & \lambda \end{bmatrix} \mathbf{f} \tag{5.13}$$

Substituting (5.11) and (5.13) into (5.10), yields

$$\begin{bmatrix} \lambda & 0 \\ 0 & \lambda \end{bmatrix} \mathbf{f} = [\mathsf{C}^{-1}]^T \int \mathsf{B}^T\mathsf{D}\mathsf{B}\; d_{\text{vol}}\, \mathsf{C}^{-1} \begin{bmatrix} \lambda & 0 \\ 0 & \lambda \end{bmatrix} \boldsymbol{\delta}$$

$$\mathbf{f} = \begin{bmatrix} \lambda^T & 0 \\ 0 & \lambda^T \end{bmatrix} [\mathsf{C}^{-1}]^T \int \mathsf{B}^T\mathsf{D}\mathsf{B}\; d_{\text{vol}}\, \mathsf{C}^{-1} \begin{bmatrix} \lambda & 0 \\ 0 & \lambda \end{bmatrix} \boldsymbol{\delta} \qquad (5.14)$$

The stiffness of the element is then

$$\mathsf{K} = \begin{bmatrix} \lambda^T & 0 \\ 0 & \lambda^T \end{bmatrix} [\mathsf{C}^{-1}]^T \int \mathsf{B}^T\mathsf{D}\mathsf{B}\; d_{\text{vol}}\, \mathsf{C}^{-1} \begin{bmatrix} \lambda & 0 \\ 0 & \lambda \end{bmatrix} \qquad (5.15)$$

The integration in (5.15) presents some difficulties if an exact solution is to be obtained. The multiplication of $\mathsf{B}^T\mathsf{D}\mathsf{B}$ would first have to be done, which would give a 6×6 matrix containing x' to various positive powers and y to powers -1 and -2. The x' terms present no difficulty but the y would have to be written in terms of x' which would lead to rather complicated expressions with x' in the denominator. To avoid this complexity, the integration may be done by the use of Gaussian Quadratures [8] [16]. This requires that $\mathsf{B}^T\mathsf{D}\mathsf{B}$ be evaluated for certain specific x' values and multiplied by weighting factors. The accumultation of $\mathsf{B}^T\mathsf{D}\mathsf{B}$ over all prescribed values of x' gives a good approximation to the integral. This requires more computer time but far less theoretical development than the closed form solution.

As each element stiffness is calculated, it can be added to the general stiffness matrix in a manner similar to that of previously discussed assemblies of elements. Thus, the general stiffness matrix is accumulated.

Loads

Only loads that have axial symmetry can be dealt with by the element under consideration. If the load is uniformly distributed along a circumferential line, a node is placed at the point where the line intersects the xy plane, and the total load on the line, broken into x and y components, is applied at the node. An edge load consisting of a bending moment can be similarly treated. Frequently in practice, the load is a uniform pressure acting on the shell surface. If the pressure load on an element is taken as distributed between the force components at the ends of the element in any proportion other than those of the consistent load vector, very poor results can be expected. To obtain the consistent load vector, the method given in Chapter 3 for the linear strain triangle can be repeated for the conical element. For constant pressure, p, the consistent load vector is, thus, found to be

$$\mathbf{F}_p = 2\pi p \begin{bmatrix} \lambda^T & 0 \\ 0 & \lambda^I \end{bmatrix} [\mathsf{C}^{-1}]^T \begin{Bmatrix} 0 \\ 0 \\ 1/2(y_1 + y_2)L \\ 1/6(y_1 + 2y_2)L^2 \\ 1/12(y_1 + 3y_2)L^3 \\ 1/20(y_1 + 4y_2)L^4 \end{Bmatrix} \quad (5.16)$$

where $[\lambda]$ is given in (5.12), C is given in (5.3), and y_1, y_2, and L are shown in Figure 5-3.

Stresses

For given loads and support systems, the displacements of the nodes can be found in the usual manner. By extracting the appropriate components from **U**, the element displacements can be obtained. These can be resolved into components along the rotated axes of Figure 5-3 by the transformation of (5.11). These, when substituted into (5.8), give strains which, when substituted into (5.6), give "stresses." These substitutions give

$$\boldsymbol{\sigma} = \mathsf{DBC}^{-1} \begin{bmatrix} \lambda & 0 \\ 0 & \lambda \end{bmatrix} \boldsymbol{\delta} \quad (5.17)$$

True stresses can then be found from (5.5).

It will be noted that, in B, of (5.17), there are variable terms, which means that stress varies with location in the element. Hence, a point must be selected in the element at which stresses are to be calculated before (5.17) can be evaluated.

A Program to Determine Stresses in an Axi-Symmetric Shell under Uniform Pressure and Circumferential Line Loads

Program AC22B uses the processes described in this chapter to solve shell problems. The load may consist of either uniform pressure or circumferential line loads or both. However, the shell and all loads must have axial symmetry, and the shell thickness must be constant. The program can best be studied by reference to Figure 5-5, AC22B Flow Chart; Figure 5-6,

```
1  PRINT TITLE
   READ AND PRINT CASE TITLE
   READ: Y.M., P.R., THICKNESS AND PRESSURE
   SET GAUSS' CONSTANTS AND FILL IN D                                GD02B
   PRINT INPUT DATA
   READ COORDINATES OF NODES                                         RC03B
   ZERO F AND S
   SET CONSTANTS IN B                                                BC01B
   REWIND 1
   DO FOR ALL ELEMENTS
     FIND COORDINATES OF END NODES: X1, X2, Y1, Y2
                        | λ 0 |
     CALCULATE AM (=    |     | ) AND FL(=ELEM LENGTH)               GL01B
                        | 0 λ |
     FILL IN C                                                       CC01B
     INVERT C                                                        IN12B
     WRITE C, AM, X1, X2, Y1, Y2 AND FL ON 1
     CALCULATE P(=∫B^T DB dv)                                        CP01B
     CALCULATE ELEM. STIFFNESS, E = AM^T C^-1T PC^-1 AM              ES02B
     ADD E TO S                                                      AS02B
   END FILE 1, REWIND 1 AND 4
   WRITE S ON 4
   END FILE 4, REWIND 4
   READ KNOWN LINE FORCES                                            KF02B
   READ KNOWN DISPLACEMENTS                                          KU01B
   SOLVE FOR UNKNOWN COMPONENTS IN U                                 GE02B
   READ S FROM 4
   SOLVE FOR F PRINT F AND U                                         FU10B
   DO FOR ALL ELEMENTS
     FIND DISPLACEMENTS AT END NODES
     READ C, AM, X1, X2, Y1, Y2, FL FROM 1
     DO FOR EACH END OF ELEMENT
       FILL IN B                                                     BV01B
       CALCULATE STRESSES                                            SS01B
   READ NEXT
             NEXT:1
   <            =            >
   CALL EXIT    GO TO 1      GO TO 1
```

Figure 5-5. AC22B Flow Chart

AC22B Fortran Statements; Figure 5-7, Instructions for AC22B Data Deck Preparation; and the computer output as given in Figure 5-10.

The program is somewhat restricted in its application but some of the limitations could easily be removed by making simple revisions to the program. For example, a variable thickness shell could be treated by dealing with a number of elements each with a different thickness. The program could be modified to read the thickness of each element and use it

```
C       *****      MAIN AC22B      *****      JUNE 29 1972
C AXI-SYMMETRIC SHELL UNDER PRESSURE (USING CONSISTENT LOAD VECTOR)  AND
C       CIRCUMFERENTIAL LINE LOAD                W H BOWES
      DIMENSION A(6),H(6),D(4,4),E(6,6),B(4,6),C(6,6),AM(6,6),P(6,6)
     1,UE(6),STR(4),NAME(20)
      DIMENSION S(6,120),U(120),F(120),IFX(120),XY(2,60)
      LBAND=6
      NBAND=6
    1 WRITE(3,102)
      READ(2,100)NAME
      WRITE(3,101)NAME
      READ(2,103) YM,GNU,T,PR
      CALL GD02B(YM,GNU,T,A,H,D)
      WRITE(3,106) PR
      PR=2.*3.14159*PR
      CALL RC03B(XY,NE,NDF)
      DO 10 J=1,NDF
      F(J)=0.
      DO 10 I=1,NBAND
   10 S(I,J)=0.
      CALL BC01B(B)
      REWIND 1
      DO 20 KE=1,NE
      X1=XY(1,KE)
      X2=XY(1,KE+1)
      Y1=XY(2,KE)
      Y2=XY(2,KE+1)
      CALL GL01B(X1,X2,Y1,Y2,AM,FL)
      CALL CC01B(C,FL)
      CALL IN12B(C,6)
      WRITE(1) C,AM,X1,X2,Y1,Y2,FL
      CALL FP01B(PR,Y1,Y2,C,FL,AM,KE,F)
      CALL CP01B(D,B,X1,X2,Y1,Y2,FL,A,H,P)
      CALL ES02B(P,C,AM,E)
   20 CALL AS02B(E,S,KE,LBAND)
      END FILE 1
      REWIND 1
      REWIND 4
      DO 25 I=1,NBAND
   25 WRITE(4) (S(I,J),J=1,NDF)
      END FILE 4
      REWIND 4
      CALL KF02B(F)
      CALL KU01B(U,IFX,NDF)
      CALL GE02B(F,J,S,IFX,NBAND,NDF,LBAND)
      DO 30 I=1,NBAND
   30 READ(4) (S(I,J),J=1,NDF)
      CALL FU10B(S,J,NBAND,NDF,3,LBAND)
      WRITE(3,112)
      DO 45 KE=1,NE
      DO 41 I=1,6
      II=3*(KE-1)+I
   41 UE(I)=U(II)
      READ(1) C,AM,X1,X2,Y1,Y2,FL
      DO 45 I=1,2
      GO TO (42,43),I
   42 Z=0.
      KN=KE
      GO TO 44
   43 Z=1.
      KN=KE+1
   44 CALL BV01B(B,X1,X2,Y1,Y2,FL,Z)
      CALL SS01B(D,B,C,AM,JE,T,STR)
   45 WRITE(3,114)KE,KN,STR
      READ(2,110) NEXT
      IF(NEXT-1)60,1,1
   60 CALL EXIT
  100 FORMAT(20A4)
  101 FORMAT('0CASE TITLE ---   ',20A4)
  102 FORMAT('1              MAIN AC22B      JUNE 29 1972',/,' AXI-SYMMETRIC SH
     1ELL UNDER PRESSURE LOAD AND CIRCUMFERENTIAL LINE LOAD')
  103 FORMAT(4F10.5)
  106 FORMAT(' INTERNAL PRESSURE =',F6.1,' P S I')
  110 FORMAT(I5)
  112 FORMAT('0',22X,'MEMBRANE STRESSES        BENDING STRESSES',/,'ELEM.
     1NO.  NODE NO.     MERID         CIRC          MERID          CIRC')
  114 FORMAT(I5,I10,2X,4E12.3)
      END
```

Figure 5-6. AC22B Fortran Statements

Data Deck:

A card containing the Case Title.

A card containing properties of the shell; Young's modulus, Poisson's Ratio, Thickness and Fluid Pressure. Format: 4F10.5

One card for each node containing;

Node number, x coordinate, y coordinate. Format: I5, 2F10.5

A blank card to indicate end of node data.

One card for each line load component containing;

Component number, Magnitude of force. Format: I5, F10.5

A blank card to indicate end of line load data.

One card for each known displacement component containing;

Component number, Displacement. Format: I5, F10.5

A blank card to indicate end of displacement data.

A card containing NEXT. Format: I5

NEXT = 0; End of job.

NEXT = 1; Execute a new case. Follow by data cards prepared according to all above instructions.

Figure 5-7. Instruction for AC22B Data Deck Preparation

in all calculations pertaining to that element. Similarly, variable pressure load could be treated; although, the program, as written, assumes the same pressure is applied to all elements.

The node data is presented in the usual manner with the additional requirement that the nodes must be numbered sequentially, starting at an edge of the shell. The program automatically numbers the elements in the same order, so that element 1 is taken to run from node 1 to node 2, etc.

Fluid pressure is taken as acting in the direction of the y' axis, as shown in Figure 5-3. If the pressure acts in the opposite direction, it should be punched as a negative number. Concentrated loads that act on a circumferential line are broken into components which are applied as nodal loads. The program is arranged to handle both types of load simultaneously. When only one type of load exists, which is the most common case, the other load is merely treated in the input data as a load of zero magnitude.

Stresses in pounds per square inch are calculated and printed. They are left as membrane and bending stresses, that is, those given by equation (5.5b) and those given by equation (5.5a). Bending stress was assumed to vary linearly through the thickness from compression on one surface to tension on the other. The value that is printed is the maximum stress and the sign is that which is applicable to the fibres at the surface on the $-y'$ side as shown in Figure 5-3. For total stress, the user must combine the membrane and bending stresses.

In the printed output, the direction of "circ" stresses is normal to the plane of Figure 5-3 while "merid" stresses are in the x' direction.

In most practical applications, it is necessary to locate a node on the x axis. Because of approximations in the method, stresses calculated for this node are very unreliable. One way to get good stress values at a point on the x axis is to approach it by small elements, disregard the calculated stresses near the point, and use those from further away to extrapolate up to the point.

Example

A pressure vessel with a hemispherical head has been selected to show the results that may be obtained, using the Finite Element Method, for axi-symmetric shells under the action of axi-symmetric loads. The pressure vessel is shown in Figure 5-8 and is loaded with an internal pressure of 500 psi.

Problem

For the pressure vessel shown in Figure 5-8, determine the membrane and bending stresses and compare them with the corresponding theoretical values.

Solution. A Finite Element Method solution to this problem may be obtained using the shell elements developed in this chapter. It is first necessary to subdivide the pressure vessel into a number of shell elements, using smaller elements in regions where rapid changes in the stress values may be expected. Such a subdivision of the pressure vessel, Figure 5-8, is shown in Figure 5-9. At the junction between the hemispherical head and the cylindrical section of the vessel, changes in the circumferential membrane stress and the bending stresses may be expected; and, therefore, a larger number of smaller elements were used in this region, as compared to the main cylindrical section of the pressure vessel. Similarly, in the apex region of the pressure vessel head, a large number of elements were used.

In order to reduce the number of elements, and thus the computer space and the computing time required, the symmetrical nature of the problem was utilized and only one-half of the shell analyzed. The elements were numbered from the extreme left end and a total of 39 elements were used.

The boundary conditions shown in Figure 5-9 permit a radial expansion of the cylindrical section of the pressure vessel and axial motion at the apex of the hemispherical head. Because of symmetry, no rotation would occur at these points and this is also reflected in the boundary conditions specified.

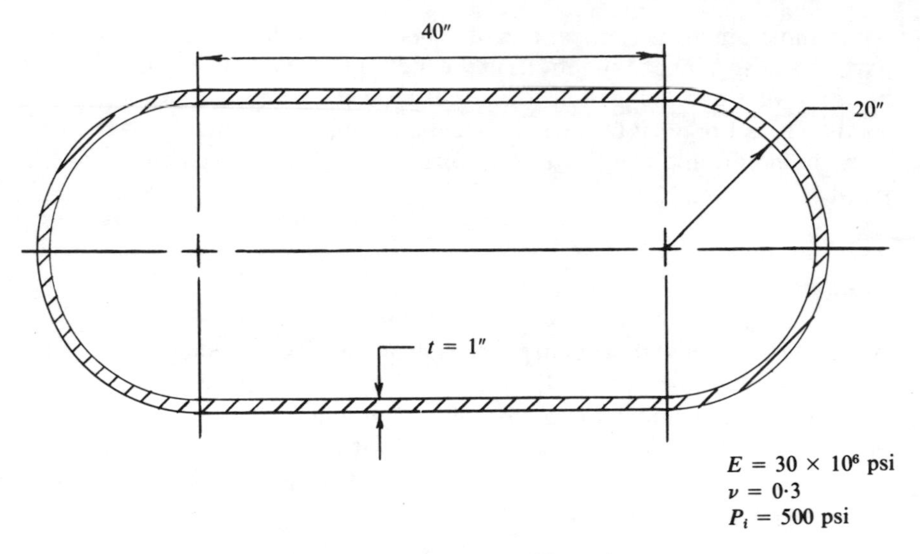

Figure 5-8. Pressure Vessel

The loading for this problem, being a uniform internal pressure, is read into the program initially and does not appear as nodal point loads.

Results

The computer output for this example is shown in Figure 5-10, and is interpreted graphically in Figures 5-11 and 5-12.

The circumferential membrane stress as found by the Finite Element Method together with the elementary membrane stress and the theoretical value of the membrane stress, based on the analysis given in reference [4] is shown in Figure 5-11. The elementary theory suggests a sudden change in the stress at the junction between the cylindrical section and the head of the vessel, while the more accurate theory in [4] gives a smooth transition of stress from the cylindrical section to the head of the vessel. The Finite Element Method values are also plotted in this figure and fall essentially on top of the theoretically correct curve.

The meridian membrane stresses as found by the Finite Element Method are given in the computer output; the maximum error in these values when compared to the theoretically correct value of 5000 psi is 2.4 percent.

The bending stresses are calculated separately in the computer program and, in Figure 5-12, the meridian bending stress as found by the Finite Element Method is compared to the theoretical values [4]. In this figure, the

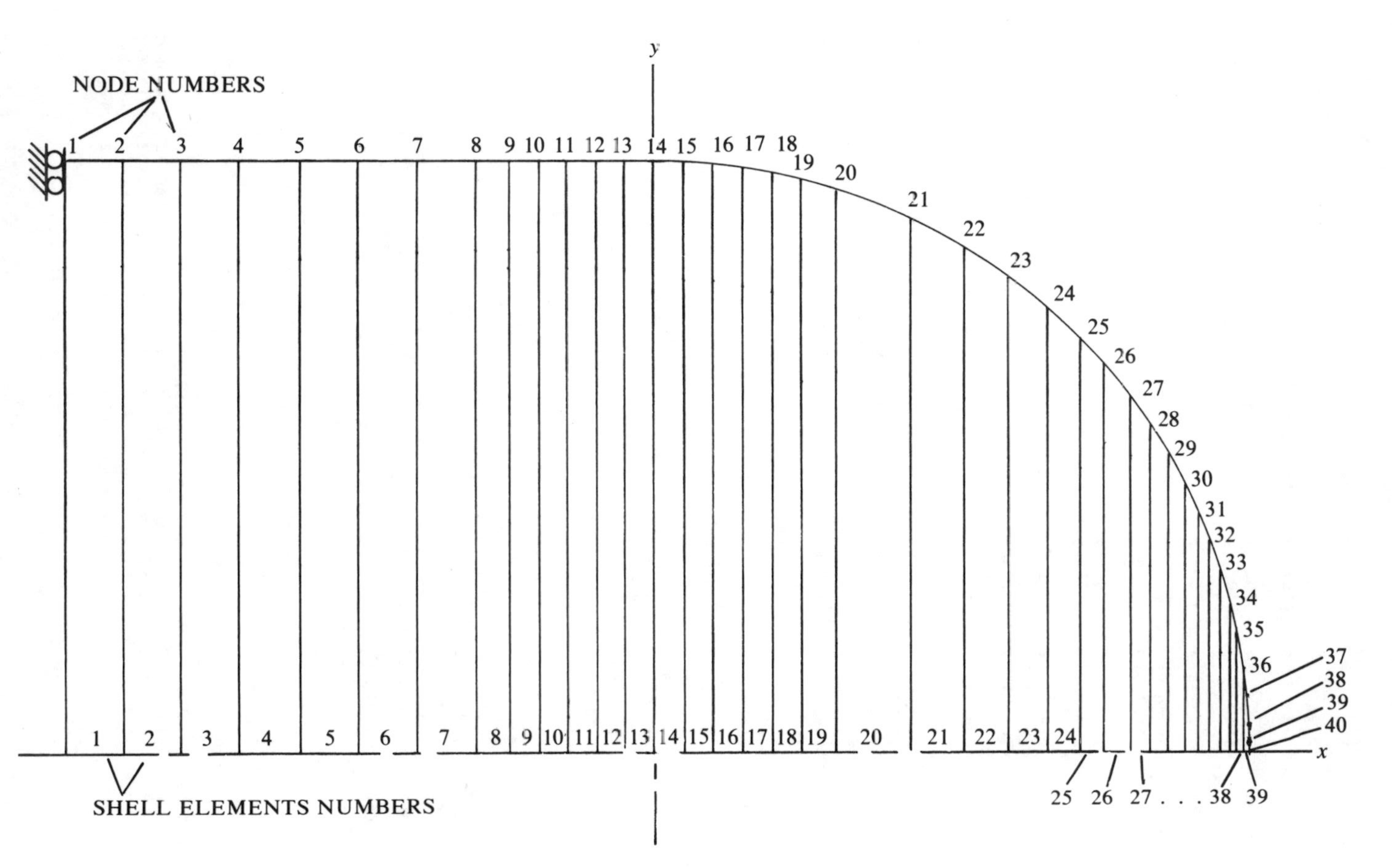

Figure 5-9. Element Subdivision for Pressure Vessel

```
          MAIN AC22B     JUNE 29 1972
AXI-SYMMETRIC SHELL UNDER PRESSURE LOAD AND CIRCUMFERENTIAL LINE LOAD

CASE TITLE ---     SPHERICAL HEAD SHELL  500 PSI

YOUNGS MODULUS     =   .30E+08
POISSONS RATIO     =  .300
THICKNESS          = 1.000
INTERNAL PRESSURE  = 500.0 P S I

NODE NO.     X-COORD       Y-COORD
    1      -.2000E+02    .2000E+02
    2      -.1800E+02    .2000E+02
    3      -.1600E+02    .2000E+02
    4      -.1400E+02    .2000E+02
    5      -.1200E+02    .2000E+02
    6      -.1000E+02    .2000E+02
    7      -.8000E+01    .2000E+02
    8      -.6000E+01    .2000E+02
    9      -.5000E+01    .2000E+02
   10      -.4000E+01    .2000E+02
   11      -.3000E+01    .2000E+02
   12      -.2000E+01    .2000E+02
   13      -.1000E+01    .2000E+02
   14       .0000E+00    .2000E+02
   15       .1000E+01    .1997E+02
   16       .2000E+01    .1990E+02
   17       .3000E+01    .1977E+02
   18       .4000E+01    .1960E+02
   19       .5000E+01    .1936E+02
   20       .6245E+01    .1900E+02
   21       .8717E+01    .1800E+02
   22       .1054E+02    .1700E+02
   23       .1200E+02    .1600E+02
   24       .1323E+02    .1500E+02
   25       .1428E+02    .1400E+02
   26       .1520E+02    .1300E+02
   27       .1600E+02    .1200E+02
   28       .1670E+02    .1100E+02
   29       .1732E+02    .1000E+02
   30       .1786E+02    .9000E+01
   31       .1833E+02    .8000E+01
   32       .1873E+02    .7000E+01
   33       .1908E+02    .6000E+01
   34       .1936E+02    .5000E+01
   35       .1960E+02    .4000E+01
   36       .1977E+02    .3000E+01
   37       .1990E+02    .2000E+01
   38       .1997E+02    .1000E+01
   39       .1999E+02    .5000E+00
   40       .2000E+02    .0000E+00

  KNOWN NON-ZERO LOADS
COMPONENT NUMBER    LOAD

      KNOWN DISPLACEMENTS
COMPONENT NUMBER    DISPLACEMENT
            1         .0000E+00
            3         .0000E+00
          119         .0000E+00
          120         .0000E+00

NODE NO.      FORCE AND DISPLACEMENT COMPONENTS
    1      -.6283E+06   .6283E+05   .2062E+05   .3033E-15   .5658E-02  -.2984E-16
    2      -.5466E-10   .1257E+06  -.1180E-09   .1336E-03   .5660E-02   .2180E-05
    3      -.1830E-09   .1257E+06   .4741E-10   .2670E-03   .5668E-02   .6134E-05
    4      -.5282E-09   .1257E+06  -.1941E-09   .4001E-03   .5686E-02   .1256E-04
```

Figure 5-10. Output Processed by AC22B

5	-,1063E-09	,1257E+06	-,9850E-09	,5324E-03	,5718E-02	,1930E-04
6	-,3294E-09	,1257E+06	-,1088E-08	,6636E-03	,5758E-02	,1867E-04
7	-,5829E-09	,1257E+06	-,1291E-08	,7938E-03	,5778E-02	-,4919E-05
8	-,1237E-08	,9425E+05	-,1571E+05	,9245E-03	,5709E-02	-,7367E-04
9	-,1223E-08	,6283E+05	-,5063E-08	,9912E-03	,5607E-02	-,1311E-03
10	-,1796E-08	,6283E+05	-,5307E-08	,1060E-02	,5441E-02	-,2043E-03
11	-,9960E-09	,6283E+05	-,6637E-08	,1132E-02	,5195E-02	-,2894E-03
12	-,7348E-09	,6283E+05	-,5167E-08	,1208E-02	,4861E-02	-,3768E-03
13	-,1824E-08	,6283E+05	-,2151E-08	,1290E-02	,4446E-02	-,4493E-03
14	,7851E+03	,6282E+05	,6529E+00	,1378E-02	,3977E-02	-,4790E-03
15	,3167E+04	,6273E+05	,1631E+02	,1462E-02	,3505E-02	-,4493E-03
16	,6278E+04	,6249E+05	,3001E+02	,1533E-02	,3082E-02	-,3763E-03
17	,9404E+04	,6210E+05	,5061E+02	,1599E-02	,2737E-02	-,2853E-03
18	,1258E+05	,6154E+05	,6650E+02	,1669E-02	,2479E-02	-,1918E-03
19	,1809E+05	,6818E+05	,3104E+04	,1749E-02	,2305E-02	-,1052E-03
20	,4033E+05	,1100E+06	,2619E+05	,1872E-02	,2188E-02	-,1813E-04
21	,5655E+05	,1216E+06	-,1440E+05	,2155E-02	,2057E-02	-,6289E-05
22	,5341E+05	,8783E+05	-,5967E+04	,2368E-02	,1943E-02	,3432E-05
23	,5027E+05	,6779E+05	-,3243E+04	,2546E-02	,1836E-02	,7059E-05
24	,4712E+05	,5387E+05	-,2053E+04	,2698E-02	,1728E-02	,7170E-05
25	,4398E+05	,4339E+05	-,1410E+04	,2828E-02	,1619E-02	,6121E-05
26	,4084E+05	,3512E+05	-,1037E+04	,2941E-02	,1508E-02	,4940E-05
27	,3770E+05	,2840E+05	-,7914E+03	,3039E-02	,1395E-02	,3932E-05
28	,3456E+05	,2285E+05	-,6280E+03	,3124E-02	,1281E-02	,2829E-05
29	,3142E+05	,1821E+05	-,5131E+03	,3199E-02	,1165E-02	,1693E-05
30	,2827E+05	,1431E+05	-,4297E+03	,3263E-02	,1050E-02	,6175E-06
31	,2513E+05	,1103E+05	-,3689E+03	,3318E-02	,9331E-03	-,8106E-07
32	,2199E+05	,8254E+04	-,3239E+03	,3365E-02	,8165E-03	-,1328E-06
33	,1885E+05	,5964E+04	-,2858E+03	,3405E-02	,6999E-03	-,5627E-06
34	,1571E+05	,4095E+04	-,2616E+03	,3438E-02	,5832E-03	-,1050E-05
35	,1257E+05	,2595E+04	-,2410E+03	,3464E-02	,4664E-03	-,8372E-06
36	,9425E+04	,1453E+04	-,2270E+03	,3484E-02	,3498E-03	-,5068E-06
37	,6283E+04	,6546E+03	-,2168E+03	,3498E-02	,2332E-03	-,1056E-05
38	,2710E+04	,1806E+03	-,3162E+03	,3505E-02	,1166E-03	-,3128E-05
39	,7854E+03	,2270E+02	-,2624E+02	,3506E-02	,5826E-04	-,1590E-05
40	,1178E+03	,1462E+01	-,9819E+01	,3506E-02	-,1961E-20	,1996E-18

		MEMBRANE STRESSES		BENDING STRESSES	
ELEM.NO.	NODE NO.	MERID	CIRC	MERID	CIRC
1	1	,500E+04	,999E+04	,140E+02	,419E+01
1	2	,500E+04	,999E+04	,220E+02	,659E+01
2	2	,500E+04	,999E+04	,225E+02	,675E+01
2	3	,500E+04	,100E+05	,427E+02	,128E+02
3	3	,500E+04	,100E+05	,441E+02	,132E+02
3	4	,501E+04	,100E+05	,618E+02	,185E+02
4	4	,499E+04	,100E+05	,647E+02	,194E+02
4	5	,501E+04	,101E+05	,464E+02	,139E+02
5	5	,499E+04	,101E+05	,508E+02	,152E+02
5	6	,501E+04	,101E+05	-,611E+02	-,183E+02
6	6	,499E+04	,101E+05	-,569E+02	-,171E+02
6	7	,500E+04	,102E+05	-,332E+03	-,996E+02
7	7	,501E+04	,102E+05	-,334E+03	-,100E+03
7	8	,498E+04	,101E+05	-,800E+03	-,240E+03
8	8	,502E+04	,101E+05	-,812E+03	-,244E+03
8	9	,497E+04	,990E+04	-,108E+04	-,325E+03
9	9	,504E+04	,992E+04	-,109E+04	-,326E+03
9	10	,496E+04	,965E+04	-,133E+04	-,398E+03
10	10	,506E+04	,968E+04	-,133E+04	-,400E+03
10	11	,494E+04	,927E+04	-,147E+04	-,441E+03
11	11	,508E+04	,932E+04	-,148E+04	-,444E+03
11	12	,491E+04	,877E+04	-,140E+04	-,421E+03
12	12	,510E+04	,882E+04	-,141E+04	-,424E+03
12	13	,489E+04	,814E+04	-,973E+03	-,292E+03
13	13	,511E+04	,820E+04	-,986E+03	-,296E+03
13	14	,488E+04	,743E+04	,554E+01	,166E+01
14	14	,512E+04	,750E+04	-,853E+01	,642E+01
14	15	,489E+04	,673E+04	,995E+03	,307E+03

Figure 5-10 (continued)

```
  15        15        .511E+04      .680E+04      .981E+03      .320E+03
  15        16        .491E+04      .612E+04      .143E+04      .451E+03
  16        16        .509E+04      .617E+04      .142E+04      .461E+03
  16        17        .492E+04      .563E+04      .158E+04      .500E+03
  17        17        .506E+04      .567E+04      .157E+04      .508E+03
  17        18        .494E+04      .528E+04      .149E+04      .472E+03
  18        18        .503E+04      .531E+04      .148E+04      .478E+03
  18        19        .496E+04      .506E+04      .131E+04      .413E+03
  19        19        .501E+04      .507E+04      .123E+04      .393E+03
  19        20        .497E+04      .495E+04      .989E+03      .301E+03
  20        20        .498E+04      .495E+04      .320E+03      .101E+03
  20        21        .497E+04      .492E+04     -.171E+03     -.493E+02
  21        21        .498E+04      .492E+04      .204E+03      .638E+02
  21        22        .498E+04      .492E+04     -.497E+02     -.164E+02
  22        22        .498E+04      .492E+04      .114E+03      .326E+02
  22        23        .499E+04      .494E+04     -.488E+02     -.184E+02
  23        23        .499E+04      .494E+04      .449E+02      .928E+01
  23        24        .499E+04      .495E+04     -.454E+02     -.182E+02
  24        24        .499E+04      .495E+04      .169E+02      .149E+00
  24        25        .499E+04      .497E+04     -.439E+02     -.177E+02
  25        25        .499E+04      .497E+04      .116E+01     -.449E+01
  25        26        .500E+04      .498E+04     -.328E+02     -.141E+02
  26        26        .499E+04      .498E+04      .227E+01     -.377E+01
  26        27        .500E+04      .499E+04     -.310E+02     -.131E+02
  27        27        .499E+04      .499E+04     -.252E+01     -.478E+01
  27        28        .500E+04      .499E+04     -.296E+02     -.120E+02
  28        28        .500E+04      .499E+04     -.547E+01     -.492E+01
  28        29        .500E+04      .500E+04     -.282E+02     -.106E+02
  29        29        .500E+04      .500E+04     -.700E+01     -.434E+01
  29        30        .500E+04      .500E+04     -.252E+02     -.847E+01
  30        30        .500E+04      .500E+04     -.597E+01     -.272E+01
  30        31        .500E+04      .500E+04     -.151E+02     -.440E+01
  31        31        .500E+04      .500E+04      .312E+01      .108E+01
  31        32        .500E+04      .500E+04     -.457E+01     -.111E+01
  32        32        .500E+04      .500E+04      .139E+02      .443E+01
  32        33        .500E+04      .500E+04     -.268E+02     -.670E+01
  33        33        .500E+04      .500E+04     -.919E+01     -.140E+01
  33        34        .500E+04      .500E+04     -.480E+01      .159E+01
  34        34        .500E+04      .500E+04      .152E+02      .763E+01
  34        35        .500E+04      .500E+04     -.636E+01      .115E+01
  35        35        .500E+04      .500E+04      .159E+02      .785E+01
  35        36        .500E+04      .500E+04     -.330E+01      .150E+01
  36        36        .499E+04      .500E+04      .261E+02      .103E+02
  36        37        .499E+04      .500E+04     -.407E+02     -.434E+01
  37        37        .499E+04      .500E+04      .133E+01      .830E+01
  37        38        .499E+04      .499E+04     -.514E+02      .314E+02
  38        38        .499E+04      .499E+04      .694E+02      .677E+02
  38        39        .499E+04      .499E+04      .631E+02      .666E+02
  39        39        .499E+04      .499E+04      .832E+02      .727E+02
  39        40        .384E+04      .115E+04      .373E+02      .112E+02
*EXIT*
```

Figure 5-10 (continued)

values found by the Finite Element Method lie on the theoretical curve in the cylindrical region of the pressure vessel. The Finite Element values tend to be slightly larger than the theoretical values in the maximum bending stress region of the hemispherical head. These errors are not large, however, and the Finite Element Method values provide a good representation of the theoretical meridian bending stress.

The circumferential bending stress is shown, in [4], to be ν times meridian bending stress; and this relation is confirmed by the values of the circumferential bending stress shown in the computer output.

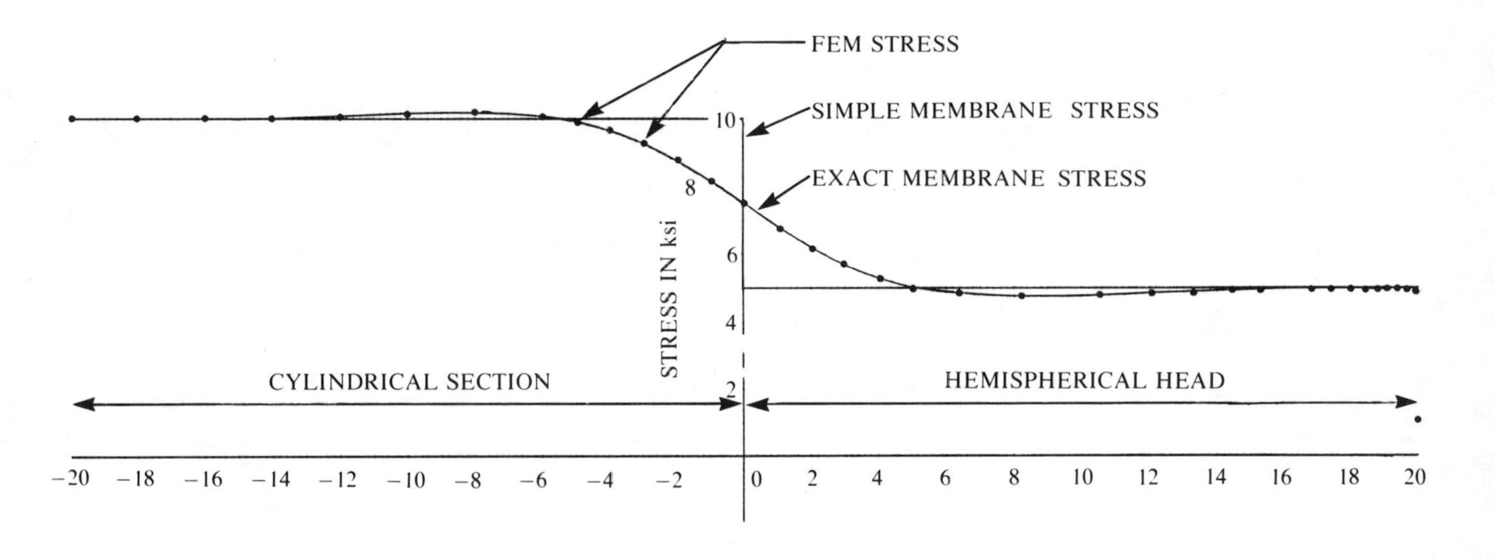

Figure 5-11. Circumferential Membrane Stress

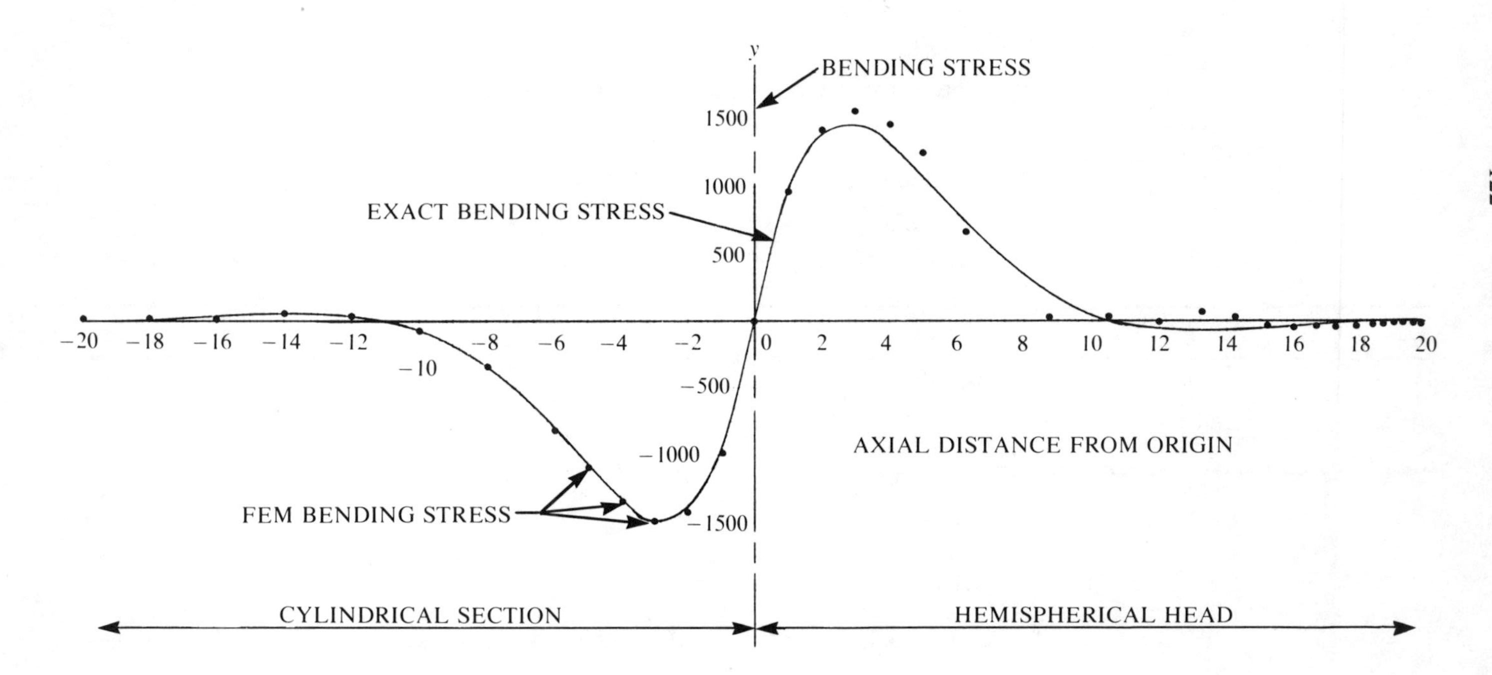

Figure 5-12. Meridional Bending Stress

The number of elements used in this illustrative example is indeed small; but, the results obtained serve to indicate that good accuracy can be obtained for the solution to many problems not amenable to a theoretical solution.

6 Axi-Symmetric Solids

The methods developed in Chapter 5 cannot be applied in cases where the shell wall is thick or where the stress concentration near a change in section is to be determined. To solve these problems and many others for cases where the solid and the loads have complete axial symmetry, we will introduce a new element. The solid might be thought of as in Figure 6-1a where the load could be either axial tension, internal pressure, a combination of these, or any other load that has axial symmetry. Ring elements of various cross-section could be used for the solid; in this analysis, a ring having a quadrilateral cross section will be treated. An element having a rectangular cross-section would be easier to analyze but it cannot be fitted into tapering regions and, hence, its use is too restricted. A typical subdivision into quadrilateral ring elements is shown in Figure 6-1, while a typical cross-section is shown in Figure 6-2.

Displacements in a Quadrilateral Ring Element

The ring element can be completely described by the x and y coordinates of the four corner nodes; but if the analysis is attempted in this system of coordinates, it is found that it is impractical to carry out some of the integrations. Since it will ultimately be necessary to resort to Gaussian quadratures, it is best at this stage to set up a system of coordinates that lend themselves to integration by that method. The system shown in Figure 6-3 makes integration practicable since the limits in either variable do not contain the other variable. Also, these limits are -1 to $+1$ which is ideal for Gaussian quadratures.

This system is also convenient for specifying the displacement functions. Consider the single component of displacement, δ_1, shown in Figure 6-4. During displacement, we would like all edges to remain straight lines for the purposes of maintaining compatibility along the edges and, hence, the displaced quadrilateral will be as shown by the broken lines in Figure 6-4. These displacements are given very simply by

$$u = (1/4)(1 - \xi)(1 - \eta)\delta_1$$

If the only displacement is δ_2 at node 2, in the x direction, compatible displacements are given by

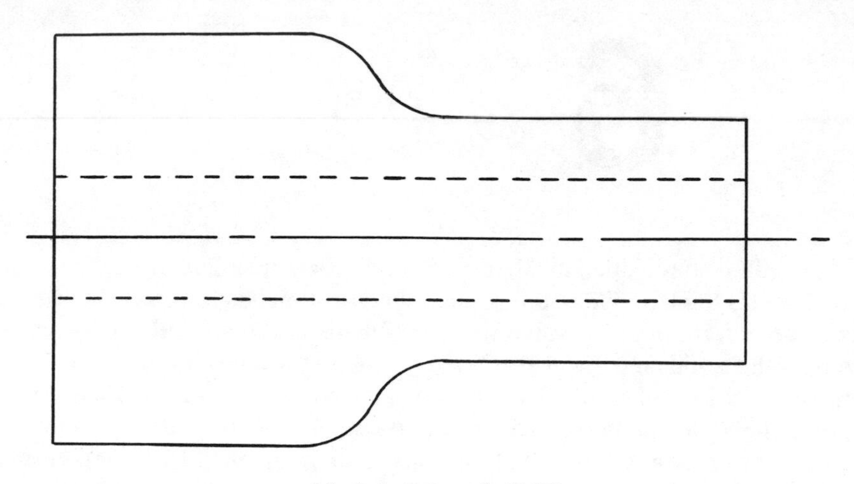

(a) Axi-symmetric Solid

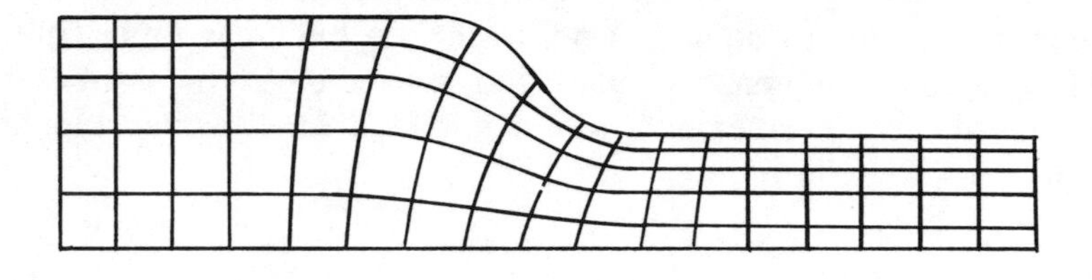

(b) Quadrilateral Element System

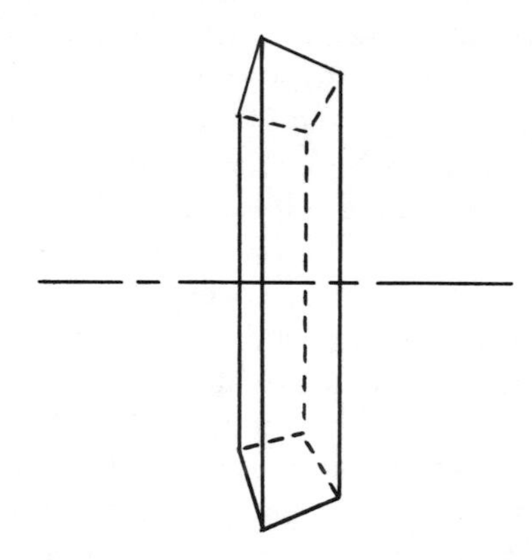

(c) Typical Quadrilateral Ring Element

Figure 6-1. Ring Elements

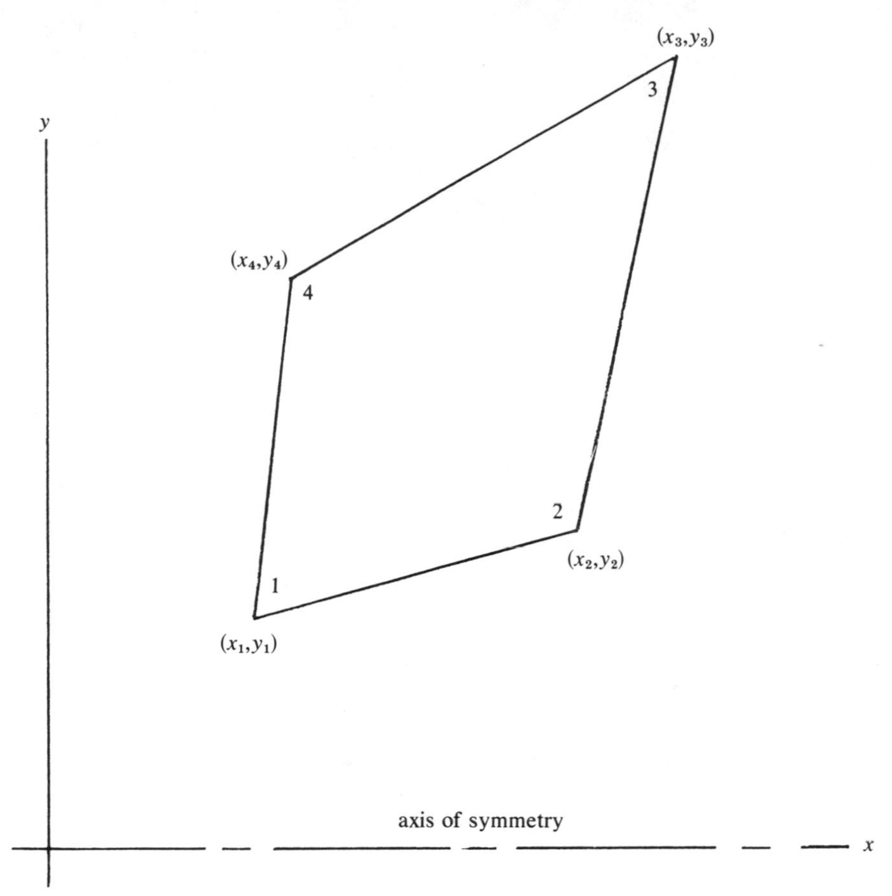

Figure 6-2. Cross-Section of a Quadrilateral Ring Element

$$u = (1/4)(1 + \xi)(1 - \eta)\delta_2$$

This may be repeated for both u and v and for all nodal displacements. By superimposing the formulas and writing in matrix form, we get displacement formulas for points in the element when all eight nodal displacements are applied simultaneously.

$$\begin{Bmatrix} u \\ v \end{Bmatrix} = \frac{1}{4}\begin{bmatrix} kl & ml & mn & kn & 0 & 0 & 0 & 0 \\ 0 & 0 & 0 & 0 & kl & ml & mn & kn \end{bmatrix} \boldsymbol{\delta} \qquad (6.1)$$

where $k = 1 - \xi$, $l = 1 - \eta$, $m = 1 + \xi$, and $n = 1 + \eta$.

These functions are referred to as interpolation functions. They also appear in the equations for transformation from $\xi\eta$ coordinates to xy coordinates, which are:

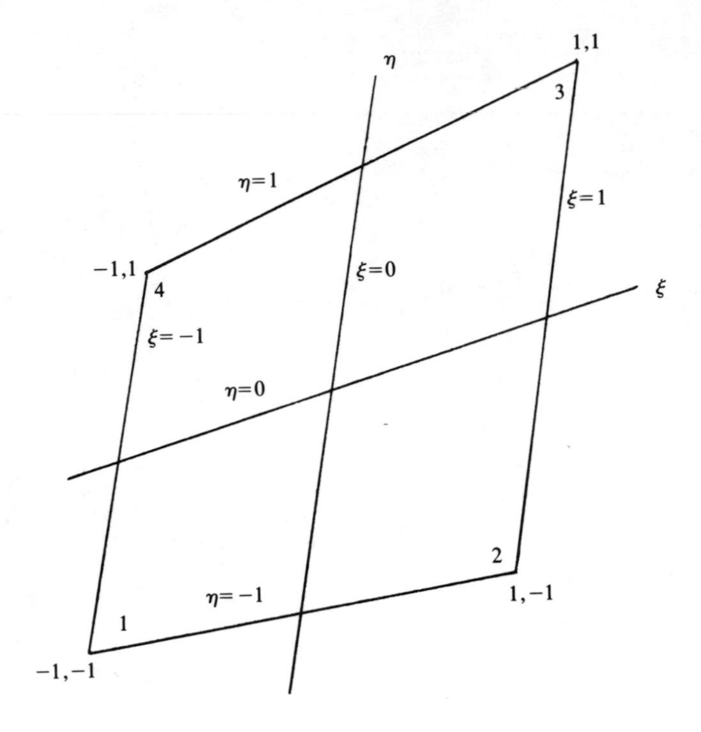

Figure 6-3. $\xi\eta$ Coordinate System

$$[\,x \quad y\,] = \frac{1}{4}\;[\,kl \quad ml \quad mn \quad kn\,] \begin{bmatrix} x_1 & y_1 \\ x_2 & y_2 \\ x_3 & y_3 \\ x_4 & y_4 \end{bmatrix} \qquad (6.2)$$

where k, l, m, and n are as above.

Stress-Strain Relationships

The stresses and strains that contribute to strain energy, and hence must be taken into account, are those in the *xy* plane and those normal to the plane.

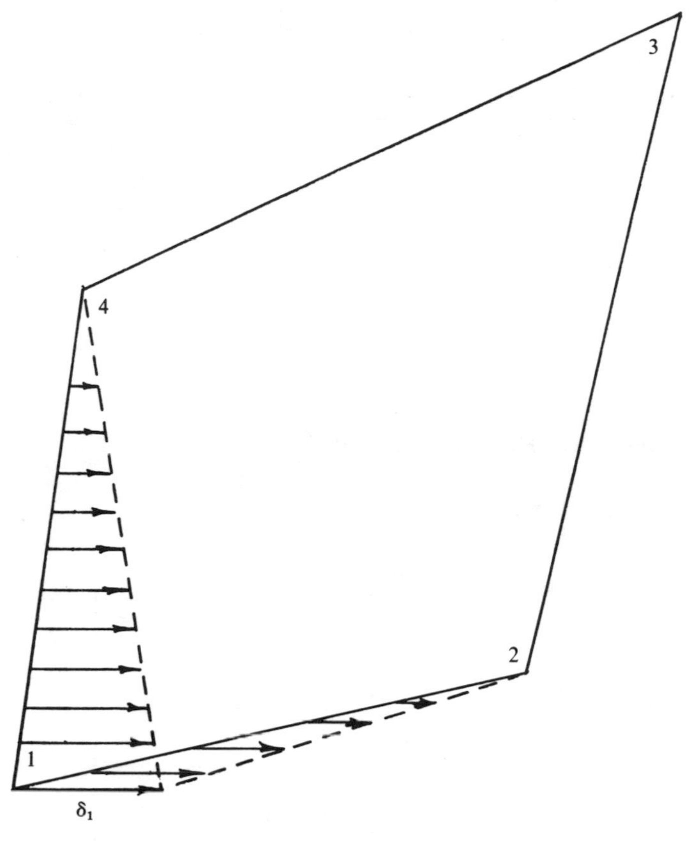

Figure 6-4. Displacements by Interpolation Functions

The latter will be indicated by the subscript θ. Thus, the stress and strain vectors are

$$\boldsymbol{\sigma} = \begin{Bmatrix} \sigma_{xx} \\ \sigma_{yy} \\ \sigma_{\theta\theta} \\ \sigma_{xy} \end{Bmatrix} \qquad \text{and} \qquad \boldsymbol{\varepsilon} = \begin{Bmatrix} \varepsilon_{xx} \\ \varepsilon_{yy} \\ \varepsilon_{\theta\theta} \\ \varepsilon_{xy} \end{Bmatrix}$$

They are related by

$$\boldsymbol{\sigma} = \mathsf{D}\boldsymbol{\varepsilon} \tag{6.3}$$

where

$$
\mathbf{D} = \frac{E(1-\nu)}{(1+\nu)(1-2\nu)} \begin{bmatrix} 1 & \frac{\nu}{1-\nu} & \frac{\nu}{1-\nu} & 0 \\ \frac{\nu}{1-\nu} & 1 & \frac{\nu}{1-\nu} & 0 \\ \frac{\nu}{1-\nu} & \frac{\nu}{1-\nu} & 1 & 0 \\ 0 & 0 & 0 & \frac{1-2\nu}{2(1-\nu)} \end{bmatrix} \tag{6.4}
$$

Strain-Displacement Relationships and Element Stiffness

The strain can be written as

$$
\boldsymbol{\epsilon} = \begin{Bmatrix} \varepsilon_{xx} \\ \varepsilon_{yy} \\ \varepsilon_{\theta\theta} \\ \varepsilon_{xy} \end{Bmatrix} = \begin{Bmatrix} \frac{\partial u}{\partial x} \\ \frac{\partial v}{\partial y} \\ \frac{v}{y} \\ \frac{\partial u}{\partial y} + \frac{\partial v}{\partial x} \end{Bmatrix} = \begin{bmatrix} \frac{\partial}{\partial x} & 0 \\ 0 & \frac{\partial}{\partial y} \\ 0 & \frac{1}{y} \\ \frac{\partial}{\partial y} & \frac{\partial}{\partial x} \end{bmatrix} \begin{Bmatrix} u \\ v \end{Bmatrix}
$$

when displacement functions (6.1) are substituted,

$$
\boldsymbol{\varepsilon} = \mathbf{B}\boldsymbol{\delta} \tag{6.5}
$$

where

$$
\mathbf{B} = \frac{1}{4} \begin{bmatrix} \frac{\partial}{\partial x} & 0 \\ 0 & \frac{\partial}{\partial y} \\ & \frac{1}{y} \\ \frac{\partial}{\partial y} & \frac{\partial}{\partial x} \end{bmatrix} \begin{bmatrix} kl & ml & mn & kn & 0 & 0 & 0 & 0 \\ 0 & 0 & 0 & 0 & kl & ml & mn & kn \end{bmatrix} \tag{6.6}
$$

The differentiations of (6.6) cannot be performed directly as ξ and η, although related to x and y through (6.2), cannot be written as explicit functions of x and y. Consequently we must resort to rather tedious manipulations to carry out the differentiations.

By the chain rule of differentiation,

$$\frac{\partial}{\partial \xi} = \frac{\partial x}{\partial \xi}\frac{\partial}{\partial x} + \frac{\partial y}{\partial \xi}\frac{\partial}{\partial y} \quad \text{and} \quad \frac{\partial}{\partial \eta} = \frac{\partial x}{\partial \eta}\frac{\partial}{\partial x} + \frac{\partial y}{\partial \eta}\frac{\partial}{\partial y}$$

or

$$\begin{Bmatrix} \dfrac{\partial}{\partial \xi} \\[2ex] \dfrac{\partial}{\partial \eta} \end{Bmatrix} = \begin{bmatrix} \dfrac{\partial x}{\partial \xi} & \dfrac{\partial y}{\partial \xi} \\[2ex] \dfrac{\partial x}{\partial \eta} & \dfrac{\partial y}{\partial \eta} \end{bmatrix} \begin{Bmatrix} \dfrac{\partial}{\partial x} \\[2ex] \dfrac{\partial}{\partial y} \end{Bmatrix} = \mathsf{J} \begin{Bmatrix} \dfrac{\partial}{\partial x} \\[2ex] \dfrac{\partial}{\partial y} \end{Bmatrix} \qquad (6.7)$$

where J is the Jacobian matrix, which can be evaluated by substitution from (6.2), giving

$$\mathsf{J} = \frac{1}{4}\begin{bmatrix} -l & l & n & -n \\ -k & -m & m & k \end{bmatrix} \begin{bmatrix} x_1 & y_1 \\ x_2 & y_2 \\ x_3 & y_3 \\ x_4 & y_4 \end{bmatrix} \qquad (6.8)$$

At any point in the element, the 2×2 J matrix can be evaluated and inverted.

Let

$$\mathsf{J}^{-1} = \bar{\mathsf{J}} = \begin{bmatrix} \bar{\mathsf{J}}_{11} & \bar{\mathsf{J}}_{12} \\ \bar{\mathsf{J}}_{21} & \bar{\mathsf{J}}_{22} \end{bmatrix}$$

then, solving (6.7), we have

$$\begin{Bmatrix} \dfrac{\partial}{\partial x} \\[2ex] \dfrac{\partial}{\partial y} \end{Bmatrix} = \mathsf{J}^{-1} \begin{Bmatrix} \dfrac{\partial}{\partial \xi} \\[2ex] \dfrac{\partial}{\partial \eta} \end{Bmatrix}$$

or

$$\frac{\partial}{\partial x} = \bar{J}_{11}\frac{\partial}{\partial \xi} + \bar{J}_{12}\frac{\partial}{\partial \eta} \quad \text{and} \quad \frac{\partial}{\partial y} = \bar{J}_{21}\frac{\partial}{\partial \xi} + \bar{J}_{22}\frac{\partial}{\partial \eta}$$

Substituting into (6.6) gives us

$$\mathbf{B} = \frac{1}{4}\begin{bmatrix} \bar{J}_{11}\frac{\partial}{\partial \xi} + \bar{J}_{12}\frac{\partial}{\partial \eta} & 0 \\ 0 & \bar{J}_{21}\frac{\partial}{\partial \xi} + \bar{J}_{22}\frac{\partial}{\partial \eta} \\ 0 & \frac{1}{y} \\ \bar{J}_{21}\frac{\partial}{\partial \xi} + \bar{J}_{22}\frac{\partial}{\partial \eta} & \bar{J}_{11}\frac{\partial}{\partial \xi} + \bar{J}_{12}\frac{\partial}{\partial \eta} \end{bmatrix}$$

$$\times \begin{bmatrix} kl & ml & mn & kn & 0 & 0 & 0 & 0 \\ 0 & 0 & 0 & 0 & kl & ml & mn & kn \end{bmatrix}$$

$$= \mathbf{B}_1 + \mathbf{B}_2 + \mathbf{B}_3 \tag{6.9}$$

where, as before, $k = 1 - \xi$, $l = 1 - \eta$, $m = 1 + \xi$, $n = 1 + \eta$, and also

$$\mathbf{B}_1 = \frac{1}{4}\begin{bmatrix} 0 & 0 \\ 0 & 0 \\ 0 & \frac{1}{y} \\ 0 & 0 \end{bmatrix}\begin{bmatrix} kl & ml & mn & kn & 0 & 0 & 0 & 0 \\ 0 & 0 & 0 & 0 & kl & ml & mn & kn \end{bmatrix} \tag{6.10a}$$

$$\mathbf{B}_2 = \frac{1}{4}\begin{bmatrix} \bar{J}_{11} & 0 \\ 0 & \bar{J}_{21} \\ 0 & 0 \\ \bar{J}_{21} & \bar{J}_{11} \end{bmatrix}\frac{\partial}{\partial \xi}\begin{bmatrix} kl & ml & mn & kn & 0 & 0 & 0 & 0 \\ 0 & 0 & 0 & 0 & kl & ml & mn & kn \end{bmatrix}$$

$$= \frac{1}{4}\begin{bmatrix} \bar{J}_{11} & 0 \\ 0 & \bar{J}_{21} \\ 0 & 0 \\ \bar{J}_{21} & \bar{J}_{11} \end{bmatrix}\begin{bmatrix} -l & l & n & -n & 0 & 0 & 0 & 0 \\ 0 & 0 & 0 & 0 & -l & l & n & -n \end{bmatrix} \tag{6.10b}$$

$$\mathsf{B}_3 = \frac{1}{4}\begin{bmatrix} \overline{\mathsf{J}}_{12} & 0 \\ 0 & \overline{\mathsf{J}}_{22} \\ 0 & 0 \\ \overline{\mathsf{J}}_{22} & \overline{\mathsf{J}}_{12} \end{bmatrix} \frac{\partial}{\partial \eta} \begin{bmatrix} kl & ml & mn & kn & 0 & 0 & 0 & 0 \\ 0 & 0 & 0 & 0 & kl & ml & mn & kn \end{bmatrix}$$

$$= \frac{1}{4}\begin{bmatrix} \overline{\mathsf{J}}_{12} & 0 \\ 0 & \overline{\mathsf{J}}_{22} \\ 0 & 0 \\ \overline{\mathsf{J}}_{22} & \overline{\mathsf{J}}_{12} \end{bmatrix} \begin{bmatrix} -k & -m & m & -k & 0 & 0 & 0 & 0 \\ 0 & 0 & 0 & 0 & -k & -m & m & k \end{bmatrix} \quad (6.10b)$$

At this stage, we are able to evaluate B at any point in the element by using equations (6.2), (6.8), (6.10), and (6.9). Thus, we establish the relationship between strain and nodal displacement through (6.5). In the general development of element stiffness in Chapter 2, this same relationship was given by equation (2.33)

$$\boldsymbol{\varepsilon} = \mathsf{BA}^{-1}\boldsymbol{\delta} \quad (2.33)$$

So that the B in this chapter is equivalent to the product BA^{-1} of Chapter 2. Using this equivalence, equation (2.38) gives the stiffness as

$$\mathsf{K} = \int \mathsf{B}^T\mathsf{DB}\, dv \quad (6.11)$$

The increment of volume is given by

$$dv = 2\pi y\, dx\, dy$$

but the independent variables are ξ and η, hence, we must eliminate dx and dy. These quantities are conveniently related through the determinant of the Jacobian matrix by

$$dx\, dy = |\mathsf{J}|\, d\xi\, d\eta$$

Hence,

$$\mathsf{K} = \int\int \mathsf{B}^T\mathsf{DB}(2\pi y)|\mathsf{J}|\, d\xi\, d\eta \quad (6.14)$$

The only practical way to integrate equation (6.14) is by some form of numerical integration. The limits on the variables were chosen to facilitate integration by Gaussian quadrature and this procedure has proved very satisfactory. As the stiffness of each element is obtained, it is added to the general stiffness matrix. When the total stiffness has been accumulated, loads can be applied and displacements found in the usual manner. From the general displacement vector, the element displacement can be extracted and, thus, stresses in any element found. Stresses vary within the

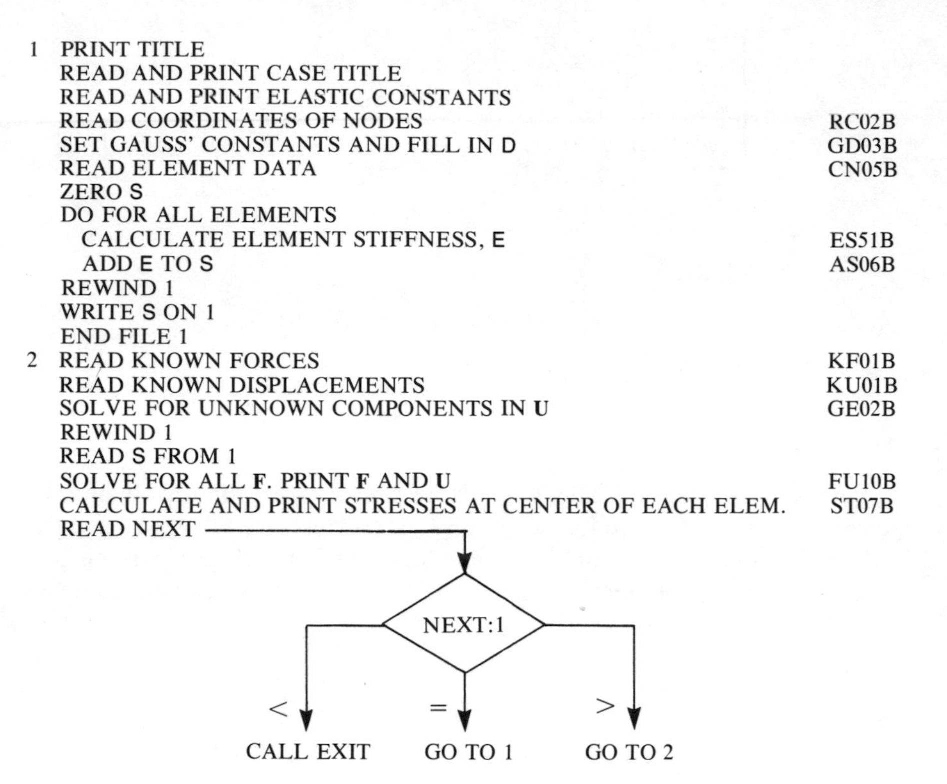

Figure 6-5. AX10B Flow Chart

element; but, as in the case of the linear strain triangle, stresses at nodal points are difficult to interpret and the most useful values are those at the origin of the $\xi\eta$ coordinate system.

A Program to Determine Stresses in an Axi-Symmetric Solid

Program AX10B uses the equations developed in this chapter to solve for stresses in solids when both the body and the loads have axial symmetry. The program is described in Figure 6-5, AX10B Flow Chart; Figure 6-6, AX10B Fortran Statements; Figure 6-7, Instructions for AX10B Data Deck Preparation; and the computer output as given in Figure 6-10.

The known force components must be calculated by the user and are those that apply to the complete circumferential line passing through the node.

```
C         *****      MAIN AX10B       *****       AUG 28 1971
C       STRESSES IN AN AXI-SYMMETRIC SOLID   (QUAD RING ELEM)   W H BOWES
      DIMENSION E(8,8),D(4,4),A(3),H(3),NAME(20),NCON(4,100)
      DIMENSION F(250),U(250),S(26,250),IFX(250),XY(2,125)
      LBAND=26
    1 WRITE(3,103)
      READ(2,100)NAME
      WRITE(3,101)NAME
      READ(2,102)YM,GNU
      WRITE(3,104)YM,GNU
      CALL RC02B(2,XY,NDF)
      CALL GD03B(YM,GNU,A,H,D)
      CALL CN05B(LBAND,NCON,NE,NBAND)
      DO 8 I=1,NBAND
      DO 8 J=1,NDF
    8 S(I,J)=0.
      DO 10 IE=1,NE
      CALL ES51B(XY,NCON,IE,D,A,H,E)
   10 CALL AS06B(IE,NCON,E,S,2,4,8,LBAND)
      REWIND 1
      DO 11 I=1,NBAND
   11 WRITE(1) (S(I,J),J=1,NDF)
      END FILE 1
    2 CALL KF01B(F,NDF)
      CALL KU01B(U,IFX,NDF)
      CALL GE02B(F,U,S,IFX,NBAND,NDF,LBAND)
      REWIND 1
      DO 12 I=1,NBAND
   12 READ(1) (S(I,J),J=1,NDF)
      CALL FU10B(S,U,NBAND,NDF,2,LBAND)
      CALL ST07B(XY,NCON,NE,D,U)
      READ(2,105)NEXT
      IF(NEXT-1)20,1,2
   20 CALL EXIT
  100 FORMAT(20A4)
  101 FORMAT('0CASE TITLE --- ',20A4)
  102 FORMAT(2F10.5,I5)
  103 FORMAT('1       MAIN AX10B    AUG 28 1971',/,'  STRESSES IN AN AXI-S
     1YMMETRIC SOLID',/,'        QUADRILATERAL RING  ELEMENTS')
  104 FORMAT('0YOUNGS MODULUS =',E10.3,/,' POISSONS RATIO =',F5.2)
  105 FORMAT(I5)
      END
```

Figure 6-6. AX10B Fortran Statements

The stresses that are printed apply at the origin of the $\xi\eta$ coordinates, that is, at the point having x and y coordinates equal to the mean of the x coordinates and y coordinates at the corner nodes of the element.

Example

This chapter deals with axi-symmetric solids, and a convenient example to use is a disc or cylinder with internal pressure. By using such an example a comparison can again be made with the theoretical results and the results given in Chapter 3.

Data Deck:

A card containing the Case Title.

A card containing Young's Modulus and Poisson's Ratio. Format: 2F10.5

One card for each node containing;

Node Number, x coordinate, y coordinate. Format: I5, 2F10.5

A blank card to indicate end of node data.

One card for each element containing;

Element Number, Node Numbers in counterclockwise order. Format: 5I5

A blank card to indicate end of element data.

One card for each known, nonzero force component containing;

Component Number, Force. Format: I5, F10.5

A blank card to indicate end of force data.

One card for each known displacement component containing;

Component Number, Displacement. Format: I5, F10.5

A blank card to indicate end of displacement data.

One card containing NEXT. Format: I5

NEXT = 0; End of job.

NEXT = 1; Execute a new case. Follow by data cards prepared according to all above instructions.

NEXT = 2; Repeat case just completed but with a new set of known forces and displacements. Follow by data cards described above starting with force data cards.

Figure 6-7. Instructions for AX10B Data Deck Preparation

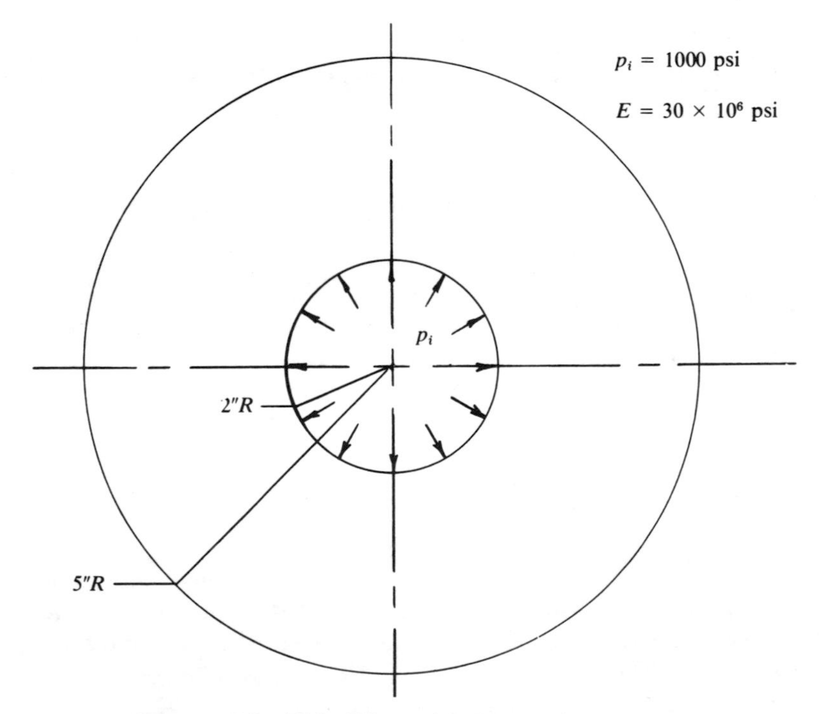

Figure 6-8. Thin Disc with Internal Pressure

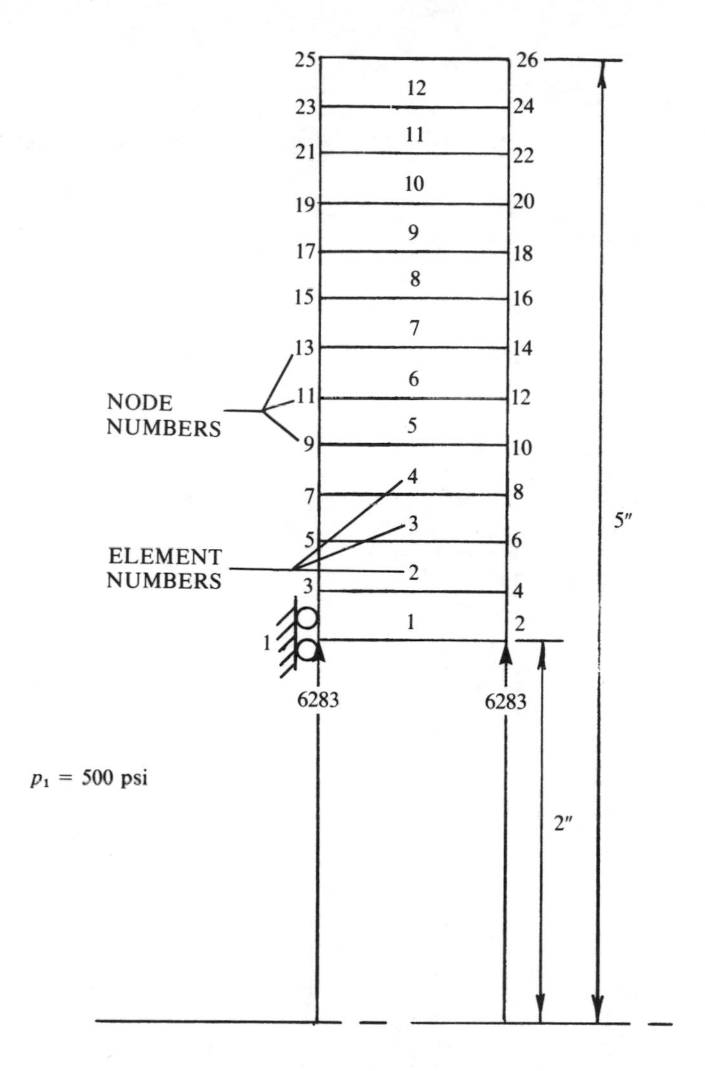

Figure 6-9. Element Subdivision for Thin Disc

Problem

For the disc shown in Figure 6-8, determine the circumferential stress by the Finite Element Method using a ring element and compare the results with the theoretical results.

Solution. The element subdivision for the disc is shown in Figure 6-9 together with the nodal loading necessary to represent the internal pressure of 500 psi.

```
        MAIN AX10B    AUG 28 1971
  STRESSES IN AN AXI-SYMMETRIC SOLID
      QUADRILATERAL RING  ELEMENTS

CASE TITLE ---    DISC UNDER THE ACTION OF 500 PSI INTERNAL PRESSURE

YOUNGS MODULUS =   .300E+08
POISSONS RATIO =   .30

NODE NO.      X-COORD         Y-COORD
     1      .0000E+00      .2000E+01
     3      .0000E+00      .2250E+01
     5      .0000E+00      .2500E+01
     7      .0000E+00      .2750E+01
     9      .0000E+00      .3000E+01
    11      .0000E+00      .3250E+01
    13      .0000E+00      .3500E+01
    15      .0000E+00      .3750E+01
    17      .0000E+00      .4000E+01
    19      .0000E+00      .4250E+01
    21      .0000E+00      .4500E+01
    23      .0000E+00      .4750E+01
    25      .0000E+00      .5000E+01
     2      .1000E+01      .2000E+01
     4      .1000E+01      .2250E+01
     6      .1000E+01      .2500E+01
     8      .1000E+01      .2750E+01
    10      .1000E+01      .3000E+01
    12      .1000E+01      .3250E+01
    14      .1000E+01      .3500E+01
    16      .1000E+01      .3750E+01
    18      .1000E+01      .4000E+01
    20      .1000E+01      .4250E+01
    22      .1000E+01      .4500E+01
    24      .1000E+01      .4750E+01
    26      .1000E+01      .5000E+01

ELEM.NO.      CORNER NODE NUMBERS
     1           1     2     4     3
     2           3     4     6     5
     3           5     6     8     7
     4           7     8    10     9
     5           9    10    12    11
     6          11    12    14    13
     7          13    14    16    15
     8          15    16    18    17
     9          17    18    20    19
    10          19    20    22    21
    11          21    22    24    23
    12          23    24    26    25

BAND WIDTH =  8
STIFFNESS CALCULATED FOR ELEM. NO.
                                1
                                2
                                3
                                4
                                5
                                6
                                7
                                8
                                9
                               10
                               11
                               12
```

Figure 6-10. Output Processed by AX10B

```
  KNOWN NON-ZERO LOADS
COMPONENT NUMBER    LOAD
        2          .6283E+04
        4          .6283E+04

     KNOWN DISPLACEMENTS
COMPONENT NUMBER   DISPLACEMENT
        1            .0000E+00

NODE NO.     FORCE AND DISPLACEMENT COMPONENTS
    1     -.7276E-11  .6283E+04  .3390E-31   .1118E-03
    2     -.5684E-13  .6283E+04 -.4429E-05   .1118E-03
    3      .3638E-11  .1546E-10 -.2771E-06   .1015E-03
    4      .2842E-11  .2274E-10 -.4152E-05   .1015E-03
    5     -.9095E-12  .1091E-10 -.3218E-06   .9348E-04
    6     -.1396E-11 -.3265E-10 -.4107E-05   .9348E-04
    7      .9095E-12  .3092E-10 -.3253E-06   .8711E-04
    8      .5146E-12  .3940E-11 -.4103E-05   .8711E-04
    9      .1000E-10  .6366E-11 -.3227E-06   .8198E-04
   10      .1609E-11 -.5016E-11 -.4106E-05   .8198E-04
   11      .1819E-11 -.1455E-10 -.3201E-06   .7780E-04
   12     -.1648E-11 -.2824E-11 -.4109E-05   .7780E-04
   13      .7276E-11  .2274E-10 -.3181E-06   .7439E-04
   14      .2270E-11  .1010E-10 -.4111E-05   .7439E-04
   15      .1819E-11 -.1273E-10 -.3167E-06   .7157E-04
   16     -.2252E-11 -.5613E-12 -.4112E-05   .7157E-04
   17      .2728E-11 -.9095E-12 -.3158E-06   .6925E-04
   18      .2046E-11  .2034E-10 -.4113E-05   .6925E-04
   19      .3638E-11  .2456E-10 -.3152E-06   .6733E-04
   20      .2487E-12  .2245E-11 -.4114E-05   .6733E-04
   21      .9095E-12  .3001E-10 -.3155E-06   .6575E-04
   22     -.5116E-12 -.8377E-11 -.4113E-05   .6575E-04
   23      .6366E-11 -.1546E-10 -.3192E-06   .6445E-04
   24     -.3784E-11 -.2574E-10 -.4110E-05   .6445E-04
   25     -.2753E-11  .2498E-11 -.3384E-06   .6339E-04
   26     -.1972E-11  .2711E-11 -.4090E-05   .6339E-04

ELEM.NO. CIRC.STR. THETA     STR1         STR2
    1    .1243E+04    .0  -.1063E+02  -.8637E+03
    2    .1035E+04   -.0  -.2802E+00  -.6534E+03
    3    .8822E+03    .0   .1126E+01  -.5003E+03
    4    .7669E+03    .0   .1041E+01  -.3853E+03
    5    .6781E+03   -.0   .7931E+00  -.2969E+03
    6    .6084E+03    .0   .5900E+00  -.2274E+03
    7    .5526E+03   -.0   .4435E+00  -.1717E+03
    8    .5073E+03    .0   .3403E+00  -.1265E+03
    9    .4700E+03   -.0   .2716E+00  -.8928E+02
   10    .4389E+03   -.0   .2495E+00  -.5824E+02
   11    .4127E+03   -.0   .3690E+00  -.3210E+02
   12    .3907E+03   -.0   .1112E+01  -.9874E+01
*EXIT*
```

Figure 6-10 (continued)

Results

The computer output is shown in Figure 6-10, and is interpreted graphically in Figure 6-11. The values for the circumferential stress fall on the theoretical curve and very little error can be detected. This analysis using 12 ring elements provides an even greater accuracy than that shown by the

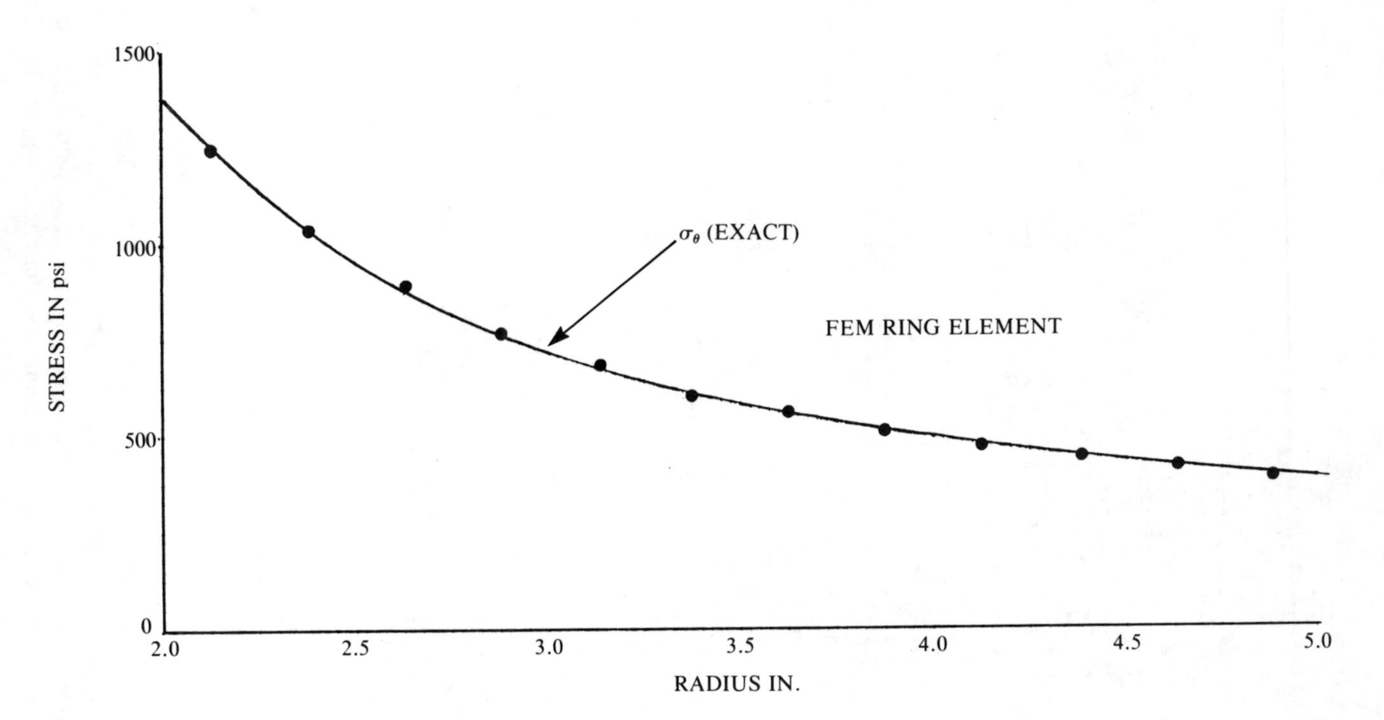

Figure 6-11. Circumferential Stresses in a Disc with Internal Pressure

analysis in Chapter 3, which utilized the linear strain triangle and the same number of elements.

7 Plates in Bending

Many practical engineering problems fall into the category of "plates in bending." Analytical solutions are available, or can be derived, for problems having complete polar symmetry and for limited classes of nonsymmetrical problems. Analysis of rectangular plates in bending frequently leads to solutions in the form of series that are difficult to evaluate. For boundaries that are neither circular nor rectangular, for plates that have varying thickness, and for nearly all plates having cut-outs, solution by analysis is impractical, but solution by the Finite Element Method is no more difficult than it is for simple cases.

Choice of Element Shape

Rectangular elements are the easiest to deal with but they give poor approximation to boundaries that are irregular. Also they do not lend themselves readily to size graduation. The extra effort required in developing the theory for triangular elements is justified by the ease with which they can be molded to fit irregular boundaries and can be graduated in size to permit small triangles in regions of large stress gradient.

Triangular elements having as few as nine degrees of freedom have been used. The theoretical development is quite simple for such elements; but the errors, because of the poor approximation to the actual displacement, are rather large. As the number of degrees of freedom is increased, the assumed displacement function may be of higher order, hence accuracy is improved but the analysis becomes more complex. A good compromise is reached at eighteen degrees of freedom where accuracy is quite adequate for engineering solutions, while the analysis is still manageable. The merits of various elements are discussed in Chapters 5 and 7 of Holand and Bell [5].

Triangular Bending Element Having 18 Degrees of Freedom

The theoretical development of the 18 degrees of freedom triangle is presented in a very clear and concise manner by Bell [5, ch. 7] and [1]. The work of Bell will not be repeated in detail here, but rather a brief outline of his procedure will be given, which should be adequate for those who wish to use the program at the end of this chapter. Anyone wishing to do further development work should read Bell carefully, but even in that case, benefit can be obtained by first reading the following description.

For deflection normal to the plane of the plate, only one displacement, w, is required. This means that instead of the displacement function being as in (2.2) it can be simplified to

$$w = [1 \;\; x \;\; y \;\; x^2 \;\; xy \;\; y^2 \; - \; - \; - \; - \; -] \, \boldsymbol{\alpha}$$

If displacements at the nodes were the only degrees of freedom used, each triangle would have three degrees of freedom and, hence, its displacement function could contain only three polynomial terms (which would be far from adequate). More degrees of freedom could be introduced by taking a number of edge nodes and this route has been followed with some success. However, we will make the main advance in the number of degrees of freedom by following the direction set in Chapter 5 where the derivative of displacement was used as a degree of freedom. Thus, at each corner node we take as degrees of freedom,

$$w, \frac{\partial w}{\partial x}, \frac{\partial w}{\partial y}, \frac{\partial^2 w}{\partial x^2}, \frac{\partial^2 w}{\partial x \, \partial y}, \text{ and } \frac{\partial^2 w}{\partial y^2}$$

It is then possible to evaluate constants for eighteen polynomial terms in the displacement function. Eighteen terms allow complete polynomials up to the fourth degree and three out of six fifth-degree terms. To include the complete family of fifth-degree terms, three more degrees of freedom are needed. These are found by introducing mid-side nodes and using the normal derivatives at these nodes as degrees of freedom. Thus an element is obtained that has complete fifth-degree polynomial displacement and twenty-one degrees of freedom.

The mid-side nodes add to the labor of defining the elements, increase the band width, and increase the total number of degrees of freedom, thus there is a strong incentive for eliminating them. This is accomplished by assuming that the variation of normal derivative on each side is a cubic. The normal derivative at the mid-side node can then be expressed in terms of the degrees of freedom at the end nodes. This enables the mid-side degree of freedom to be eliminated from the system and provides an 18×18 stiffness matrix obtained for an element having eighteen degrees of freedom.

The stiffness matrix for the assembly of triangular elements is accumulated in the usual manner. A concentrated point load at a node poses no problem as it represents a known force component with the same subscript as w at that node. Distributed load is more difficult to deal with as it must be replaced by components corresponding to all six degrees of freedom at

```
1  PRINT TITLE
   READ AND PRINT CASE TITLE
   READ AND PRINT ELASTIC CONSTANTS AND PLATE THICKNESS
   READ COORDINATES OF NODES                                        RC02B
   READ AND PRINT ELEMENT DATA INCLUDING PRESSURE                   CN08B
   DO FOR ALL ELEMENTS
     IS THIS ELEMENT SIMILAR TO ITS PREDECESSOR?
       |  \ NO                                                      AM21B
       |    \                                                       IN12B
      YES    DETERMINE ELEMENT STIFFNESS, E                         EV02B
       |    /                                                       T101B
       v  v                                                         ES18B
     DETERMINE CONSISTENT LOAD VECTOR FOR PRESSURE, PF              PF01B
     ADD E TO S AND PF TO F                                         AS03B
   READ DATA FOR LOCAL ROTATION OF AXES
   FILL IN TRANSFORMATION MATRIX FOR ROTATION                       TM02B
   ALTER S FOR ROTATED AXES                                         LR01B
   ALTER F FOR ROTATED AXES                                         LR02B
   READ NUMBERS OF NODES REQUIRING MODIFIED CURVATURES
   ALTER S FOR MODIFIED CURVATURES                                  LR01B
   ALTER F FOR MODIFIED CURVATURES                                  LR02B
   READ KNOWN FORCES                                                KF02B
   READ KNOWN DISPLACEMENTS                                         KU01B
   SOLVE FOR UNKNOWN COMPONENTS IN U                                GE02B
   CHANGE U FOR RESTORED CURVATURES                                 RU02B
   CHANGE U FOR RESTORED AXIAL DIRECTIONS                           RU02B
   PRINT DISPLACEMENTS
   CALCULATE AND PRINT MOMENTS AND STRESSES                         ST19B
   READ NEXT

       NEXT = 0 → CALL EXIT
       NEXT = 1 → GO TO 1
```

Figure 7-1. PB11B Flow Chart

each node. This is done by determining the consistent load vector, which must be found by the computer as it is impractical to evaluate by hand, by processes similar to those developed on pages 67-71.

Boundaries that are built in and those that are simply supported are quite easy to define by specifying the known displacement components at the nodes. Care must be taken to state all known components. For example, if an edge parallel to the x axis is built in, then at each node on the edge it must be specified that component numbers 1, 2, 3, 4, and 5 are zero. Whenever there is an axis of symmetry, the warping component, number 5, must be specified as zero as well as the normal slope, number 2 or 3. Some ambiguity may arise at corners where the known displacement component numbers depend on which edge the corner node is associated with. If the element at such a corner is large, the results can be significantly altered by the arbitrary choice of inputs. The best technique is to provide small elements at ambiguous corners.

```
C          *****     MAIN PB11B     *****          JUNE 25 1973
C   PLATE BENDING    TRIANGULAR ELEMENTS WITH 18 DOF                W H BOWES
      DIMENSIONS(42,150),F(150),U(150),IFX(150),XY(2,25),IROT(25),IM(25)
      DIMENSION NCON(3,40),PR(3,40)
      DIMENSION PF(18),D(3,3),A(21,21),P(7,7),G(21,18),XL(3),YL(3)
      DIMENSION X(4),Y(4),CO(3),SI(3),FL(3),NAME(20),E(18,18),CQ(21,3)
      DIMENSION T(5,5),TC(3,3),ND(4),NDL(3)
      LBAND=42
      DO 2 J=1,3
      XL(J)=0.
      YL(J)=0.
      NDL(J)=0
      DO 2 I=1,21
    2 CQ(I,J)=0.
      DO 3 KI=1,4
      DO 3 LI=1,KI
      I=(KI*(KI+3))/2+3+LI
      CQ(I,1)=(LI*(LI+1))/2
      CQ(I,2)=((KI+1-LI)*(KI+2-LI))/2
    3 CQ(I,3)=LI*(KI+1-LI)
      DO 4 I=1,5
      DO 4 J=1,5
    4 T(I,J)=0.
    1 WRITE(3,100)
      READ(2,102) NAME
      WRITE(3,104)NAME
      READ(2,106)YM,GNU,TH
      WRITE(3,108)YM,GNU,TH
      C=YM*TH**3/(12.*(1.-GNU*GNU))
      D(1,1)=C
      D(1,2)=C*GNU
      D(1,3)=0.
      D(2,1)=C*GNU
      D(2,2)=C
      D(2,3)=0.
      D(3,1)=0.
      D(3,2)=0.
      D(3,3)=C*(1.-GNU)*.5
      C=1./(1.-GNU*GNU)
      TC(1,1)=C
      TC(1,2)=0.
      TC(1,3)=-GNU*C
      TC(2,1)=0.
      TC(2,2)=1.
      TC(2,3)=0.
      TC(3,1)=-GNU*C
      TC(3,2)=0.
      TC(3,3)=C
      CALL RC02B(6,XY,NDF)
      NN=NDF/6
      CALL CN08B(LBAND,NBAND,NE,NCON,PR)
      DO 5 J=1,NDF
      F(J)=0.
      DO 5 I=1,NBAND
    5 S(I,J)=0.
      NET=0
      DO 15 IE=1,NE
      XC=0.
      YC=0.
      DO 6 I=1,3
      N=NCON(I,IE)
      ND(I)=N
      X(I)=XY(1,N)
      Y(I)=XY(2,N)
      XC=XC+X(I)/3.
    6 YC=YC+Y(I)/3.
      ND(4)=ND(1)
      DO 7 I=1,3
```

Figure 7-2. PB11B Fortran Statements

```
      ND(I)=(ND(I+1)-ND(I))/IABS(ND(I+1)-ND(I))
      X(I)=X(I)-XC
    7 Y(I)=Y(I)-YC
      X(4)=X(1)
      Y(4)=Y(1)
      DO 10 I=1,3
      XD=ABS(X(I)-XL(I))
      IF(XD-.001)8,8,12
    8 YD=ABS(Y(I)-YL(I))
      IF(YD-.001)9,9,12
    9 IF(ND(I)-NDL(I))12,10,12
   10 CONTINUE
      GO TO 14
   12 NET=NET+1
      DO 13 I=1,3
      NDL(I)=ND(I)
      XL(I)=X(I)
   13 YL(I)=Y(I)
      CALL AM21B(X,Y,A,CO,SI,FL,ND)
      CALL IN12B(A,21)
      CALL EV02B(A,CO,SI,FL,ND,G)
      CALL TI01B(7,X,Y,P)
      CALL ES18B(CQ,D,G,P,E)
   14 CALL PF01B(IE,PR,P,G,PF)
   15 CALL AS03B (IE,NCON,E,PF,S,F,6,3,18,LBAND,2)
      WRITE(3,116)NET
      REWIND 1
      IROT(1)=0
      J=1
      WRITE (3,113)
   22 READ(2,114)N,THE
      IF(N)26,26,24
   24 WRITE(3,115)N,THE
      IROT(J)=N
      J=J+1
      IROT(J)=0
      CALL TM02B(THE,T)
      WRITE(1)T
      NJ=6*N-5
      CALL LR01B(S,LBAND,NBAND,NU,T,5)
      CALL LR02B(F,NJ,T,5)
      GO TO 22
   26 END FILE 1
      REWIND 1
      IM(1)=0
      J=1
      WRITE(3,118)
   27 READ(2,112)N
      IF(N)29,29,28
   28 WRITE(3,112)N
      IM(J)=N
      J=J+1
      IM(J)=0
      NJ=6*N-3
      CALL LR01B(S,LBAND,NBAND,NU,TC,3)
      CALL LR02B(F,NU,TC,3)
      GO TO 27
   29 CALL KF02B(F)
      CALL KU01B(U,IFX,NDF)
      CALL GE02B(F,J,S,IFX,NBAND,NDF,LBAND)
      DO 31 J=1,NN
      N=IM(J)
      IF(N)35,35,30
   30 NJ=6*N-3
   31 CALL RU02B(J,NJ,TC,3)
   35 DO 37 J=1,NN
      N=IROT(J)
      IF(N)38,38,36
```

Figure 7-2 (continued)

```
36 READ(1) T
   NU=6*N-5
37 CALL RU02B(U,NU,T,5)
38 WRITE(3,119)
   DO 40 I=1,NN
   IE=6*I
   IS=IE-5
40 WRITE(3,120) I,(U(II),II=IS,IE)
   WRITE(3,110)
   CALL ST19B(D,J,NCON,TH,NN)
   READ(2,112)NEXT
   IF(NEXT-1)52,1,1
52 CALL EXIT
100 FORMAT('1            MAIN PB11B  JUNE 25 1973 ',/,' PLATE IN BENDING B
   1Y TRIANGULAR ELEMENTS (18 DEG. OF F.)')
102 FORMAT(20A4)
104 FORMAT('0CASE TITLE  ---   ',20A4)
106 FORMAT(3F10.5)
108 FORMAT('0YOUNGS MODULUS =',E10.3,'    POISSONS RATIO =',F5.3,/,
   1' PLATE THICKNESS =',F6.3)
110 FORMAT('0NODE NO.     MXX        MYY        MXY        SXX        SYY
   1     SXY        THETA      PS1        PS2')
112 FORMAT(I5)
113 FORMAT('0LOCAL ROTATED AXES',/,'NODE NO.    THETA(DEG)')
114 FORMAT(I5,F10.5)
115 FORMAT(I5,5X,F8.1)
116 FORMAT('0NUMBER OF DIFFERENT ELEMENT TYPES =',I4)
118 FORMAT('0NODES HAVING MODIFIED CURVATURES')
119 FORMAT('0NODE NO.     DISPLACEMENTS REFERRED TO GLOBAL AXES')
120 FORMAT(I6,3X,6E11.4)
    END
```

Figure 7-2 (continued)

Not all edges in real problems can be arranged to be parallel to one of the coordinate axes. To accommodate such an edge it must be possible to rotate axes locally. When there is local rotation of axes, the stiffness matrix must be altered in a manner somewhat similar to that given in Chapter 2 for guided nodes.

A free edge or one on which the bending moment is known can not be treated directly. The bending moment on the edge is determined by components 4 and 6 linked through the Poisson effect. Thus, a zero moment does not mean that either curvature is zero but rather that a combination of the two must be zero. This can be handled by a transformation which in effect makes a new component 4 equivalent to the old 4 and 6 linked through Poisson's ratio so that the new component 4 is proportional to the bending moment in the x direction. Simultaneously, curvature component 6 is also modified.

A Program to Solve Plates in Bending

Program PB11B solves plate-bending problems by the methods just described. Reference should be made to Figure 7-1, PB11B Flow Chart; Figure 7-2, PB11B Fortran Statements; Figure 7-3, Instructions for PB11B

Data Deck:

A card containing Case Title.

A card containing;

Young's Modulus, Poisson's Ratio, Plate Thickness. Format: 3F10.5

One card for each node containing;

Node number, x coordinate, y coordinate. Format: I5, 2F10.5

A blank card to indicate end of node data.

One card for each element containing;

Element number, Node numbers at corners of triangle in counterclockwise order, Pressure at centroid, Rate of pressure change in x direction, Rate of pressure change in y direction. Format: 4I5, 3F10.5

A blank card to indicate end of element data.

One card for each node having local axes rotated, containing;

Node number, angle of rotation in degrees. Format: I5, F10.5

A blank card to indicate end of rotation data.

One card for each node at which curvatures are to be modified, containing;

Node number. Format I5.

A blank card to indicate end of modified curvature data.

One card for each known, concentrated load component containing;

Component number, Load component. Format: I5, F10.5

A blank card to indicate end of concentrated load data.

One card for each known displacement component containing;

Component number, Displacement. Format: I5, F10.5

A blank card to indicate end of known displacements.

A card containing NEXT. Format I5

NEXT = 0; End of job.
NEXT = 1; Execute a new case. Follow by cards prepared according to all above instructions.

Figure 7-3. Instructions for PB11B Data Deck Preparation

Data Deck Preparation; and the computer output given for the examples at the end of the chapter.

The program is able to deal with concentrated normal forces and normal pressure that varies linearly over the surface of the element. For higher-order pressure functions, the user must approximate the pressure by specifying an equivalent pressure at the centroid and a rate of change in the coordinate directions.

For known edge moments, the user must specify the nodes at which modified curvatures are required. The axial curvatures are combined to give a modified curvature such that the modified curvature multiplied by the flexural rigidity is equal to moment intensity. Thus, for a given edge moment the modified curvature can be determined and used as a boundary condition. The results obtained by Bell, and confirmed by program PB11B, show that this method gives very good accuracy.

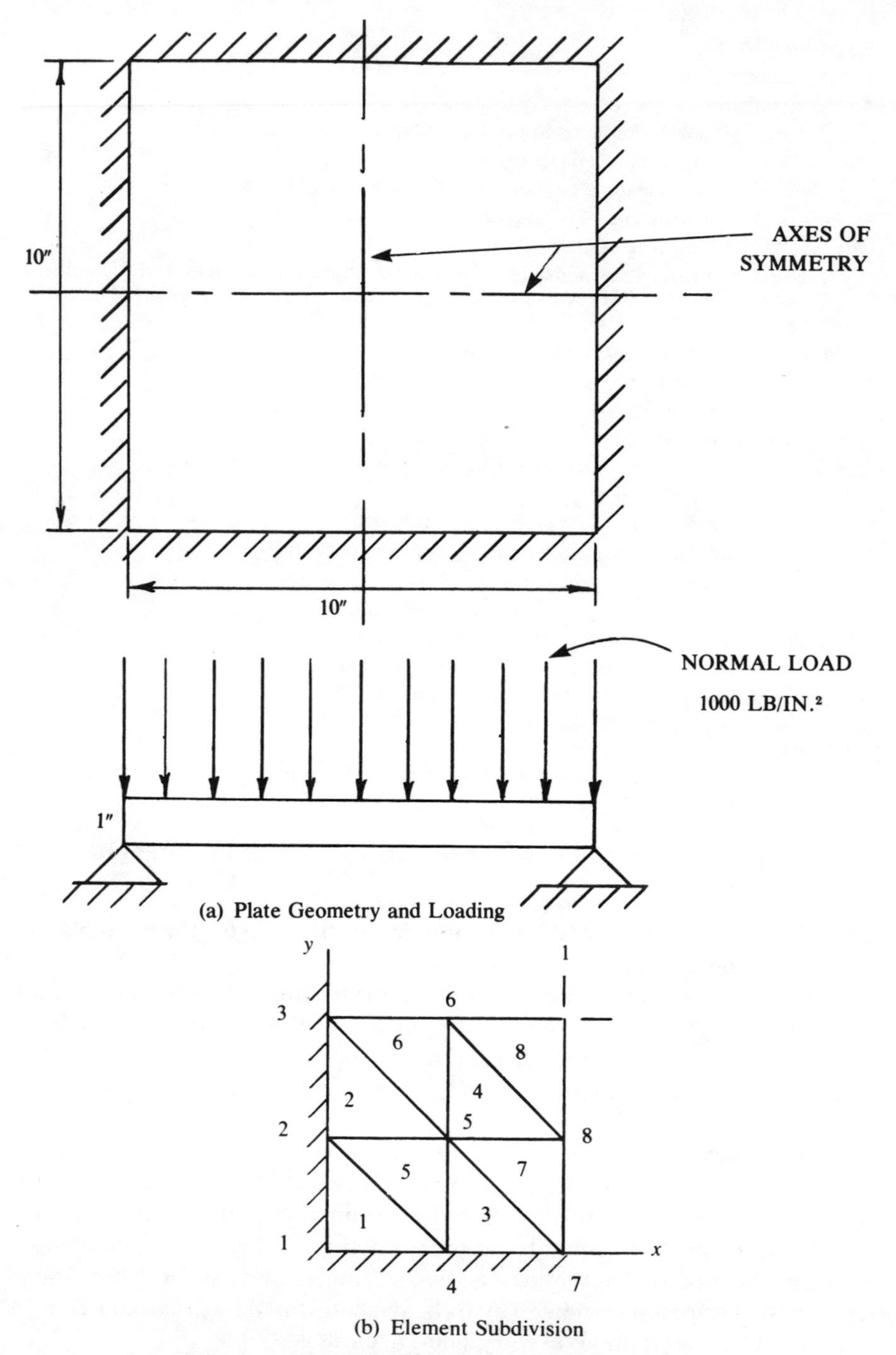

Figure 7-4. Square Plate with Normal Load

```
        MAIN PB11B  JUNE 25 1973
PLATE IN BENDING BY TRIANGULAR ELEMENTS (18 DEG. OF F.)

CASE TITLE  ---  SQUARE PLATE   UNIFORM LOAD   SIMPLY SUPPORTED

YOUNGS MODULUS =   .300E+08    POISSONS RATIO = .300
PLATE THICKNESS = 1.000

NODE NO.     X-COORD       Y-COORD
    1      .0000E+00     .0000E+00
    2      .0000E+00     .2500E+01
    3      .0000E+00     .5000E+01
    4      .2500E+01     .0000E+00
    5      .2500E+01     .2500E+01
    6      .2500E+01     .5000E+01
    7      .5000E+01     .0000E+00
    8      .5000E+01     .2500E+01
    9      .5000E+01     .5000E+01

ELEM.NO.    CORNER NODE NUMBERS    PRESSURE      DP/DX        DP/DY
   1          1      4      2     1000.000       .000         .000
   2          2      5      3     1000.000       .000         .000
   3          4      7      5     1000.000       .000         .000
   4          5      8      6     1000.000       .000         .000
   5          2      4      5     1000.000       .000         .000
   6          3      5      6     1000.000       .000         .000
   7          5      7      8     1000.000       .000         .000
   8          6      8      9     1000.000       .000         .000

BAND WIDTH = 24

NUMBER OF DIFFERENT ELEMENT TYPES =   2

 LOCAL ROTATED AXES
NODE NO.    THETA(DEG)

NODES HAVING MODIFIED CURVATURES

   KNOWN NON-ZERO LOADS
 COMPONENT NUMBER    LOAD

       KNOWN DISPLACEMENTS

COMPONENT NUMBER    DISPLACEMENT
        1            .0000E+00            20          .0000E+00
        2            .0000E+00            22          .0000E+00
        3            .0000E+00            24          .0000E+00
        4            .0000E+00            33          .0000E+00
        6            .0000E+00            35          .0000E+00
        7            .0000E+00            37          .0000E+00
        9            .0000E+00            38          .0000E+00
       10            .0000E+00            40          .0000E+00
       12            .0000E+00            41          .0000E+00
       13            .0000E+00            42          .0000E+00
       15            .0000E+00            44          .0000E+00
       16            .0000E+00            47          .0000E+00
       17            .0000E+00            50          .0000E+00
       18            .0000E+00            51          .0000E+00
       19            .0000E+00            53          .0000E+00

NODE NO.      DISPLACEMENTS REFERRED TO GLOBAL AXES
  1     -.1698E-15  .4489E-16  .4489E-16  .1784E-15  .1666E-02  .1784E-15
  2      .1450E-15  .3575E-02  .3476E-16 -.1296E-15  .1029E-02  .2139E-15
  3      .2093E-15  .4902E-02  .3751E-17 -.1027E-15  .3880E-15  .1268E-15
  4      .1450E-15  .3476E-16  .3575E-02  .2139E-15  .1029E-02 -.1296E-15
  5      .7763E-02  .2292E-02  .2292E-02 -.8282E-03  .6962E-03 -.8282E-03
```

Figure 7-5. Output Processed by PB11B: Square Plate

6	.1070E-01	.3189E-02	.3225E-15	-.1133E-02	.2091E-15	-.9533E-03
7	.2093E-15	.3751E-17	.4902E-02	.1268E-15	.3880E-15	-.1027E-15
8	.1070E-01	.3225E-15	.3189E-02	-.9533E-03	.2091E-15	-.1133E-02
9	.1479E-01	.4247E-15	.4247E-15	-.1341E-02	-.1586E-14	-.1341E-02

NODE NO.	MXX	MYY	MXY	SXX	SYY	SXY
1	.637E-09	.637E-09	.320E+04	-.382E-08	-.382E-08	-.192E+05
2	-.180E-09	.481E-09	.198E+04	.108E-08	-.289E-08	-.119E+05
3	-.178E-09	.264E-09	.746E-09	.107E-08	-.158E-08	-.448E-08
4	.481E-09	-.180E-09	.198E+04	-.289E-08	.108E-08	-.119E+05
5	-.296E+04	-.296E+04	.134E+04	.177E+05	.177E+05	-.803E+04
6	-.390E+04	-.355E+04	.402E-09	.234E+05	.213E+05	-.241E-08
7	.264E-09	-.178E-09	.746E-09	-.158E-08	.107E-08	-.448E-08
8	-.355E+04	-.390E+04	.402E-09	.213E+05	.234E+05	-.241E-08
9	-.479E+04	-.479E+04	-.305E-08	.287E+05	.287E+05	.183E-07

THETA	PS1	PS2
-45.0	.192E+05	-.192E+05
-45.0	.119E+05	-.119E+05
-36.8	.441E-08	-.493E-08
45.0	-.119E+05	.119E+05
-45.0	.258E+05	.971E+04
-.0	.234E+05	.213E+05
36.8	-.493E-08	.441E-08
.0	.213E+05	.234E+05
-45.0	.287E+05	.287E+05

Figure 7-5 (continued)

Examples

Support conditions for plates in bending have a large effect on the plate-bending stresses and, in order to properly illustrate the use of the Finite Element Method with varying supports and distribution of loads, three separate problems have been selected.

Problem 1

Determine the maximum deflection and the maximum bending stress in a ten-inch-square steel plate carrying a normal load of 1000 pounds per square inch. The plate is one inch thick and is simply supported on all edges.

Solution. As shown in Figure 7-4a the plate and load have two axes of symmetry. It is therefore necessary to solve only one-quarter of the plate. This section of the plate subdivided into 8 elements is shown in Figure 7-4b. For this problem, no local rotations and no modified curvatures are required. The boundary conditions, are interpreted as component numbers of known zero displacements and are shown in Table 7-1.

The computer output for this problem is given in Figure 7-5.

Table 7-1
Component Numbers of Known Displacements (all disp. = 0)

Node No.	w	$\frac{\partial w}{\partial x}$	$\frac{\partial w}{\partial y}$	$\frac{\partial^2 w}{\partial x^2}$	$\frac{\partial^2 w}{\partial x\,\partial y}$	$\frac{\partial^2 w}{\partial y^2}$
1	1	2	3	4		6
2	7		9	10		12
3	13		15	16	17	18
4	19	20		22		24
5						
6			33		35	
7	37	38		40	41	42
8		44			47	
9		50	51		53	

Problem 2

Determine the deflection and stresses in a vertical glass plate 1 1/4″ thick, 60″ high and 120″ long as shown in Figure 7-5. The plate is loaded by a hydrostatic pressure from water having its surface at the upper edge of the plate and the vertical edges are built-in, the upper edge is free and the bottom edge is simply supported.

Solution. The sketch of the plate in Figure 7-6 shows that only one axis of symmetry exists for both the plate and the load and hence one-half of the plate must be solved. This section of the plate is illustrated in Figure 7-7. No local rotations are necessary since the edges are parallel to the coordinate axes, however, nodes 6 and 9 are on a free edge and hence curvatures must be modified at these nodes.

Because the plate is loaded by a hydrostatic pressure, the load varies linearly from the free edge to the plate bottom but remains constant along any section parallel to the bottom of the plate. This varying load is taken into account by specifying the pressure at the centroid of each element and the rate of change of the load in each coordinate direction. This load data is given in Table 7-2.

The boundary conditions must be carefully determined. All component numbers with zero displacement are given in Table 7-3.

The computer output for this example is given in Figure 7-8.

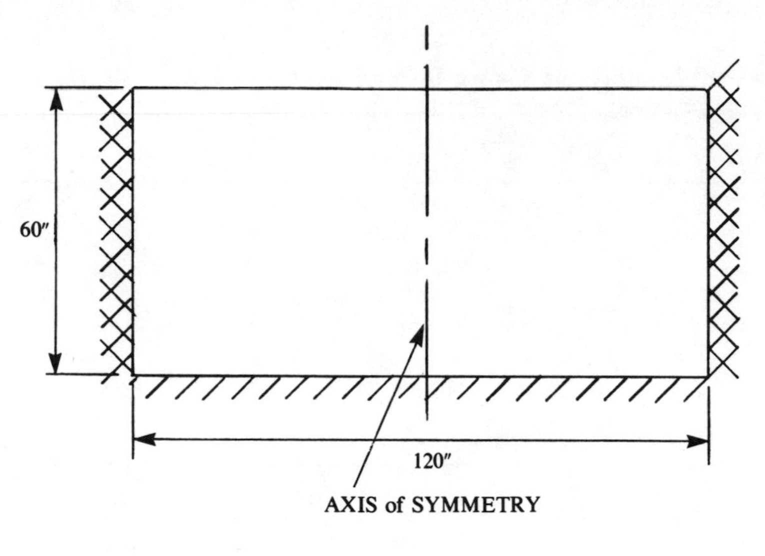

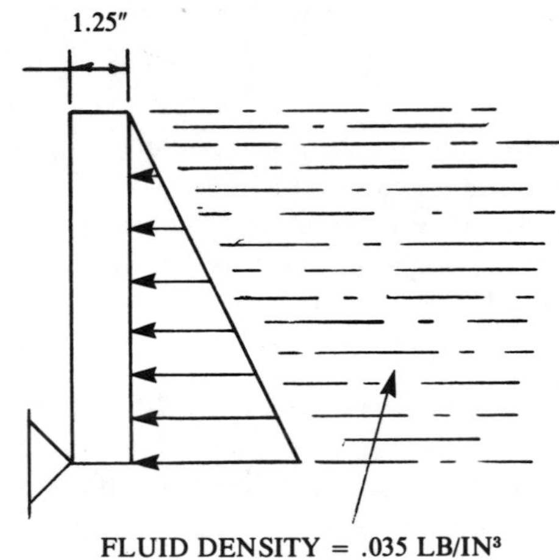

Figure 7-6. Plate with Hydrostatic Load

Problem 3

Determine the deflection and stresses in an equilateral triangular plate, one-inch thick, carrying a uniform pressure load of 1000 pounds per square inch and simply supported on all edges.

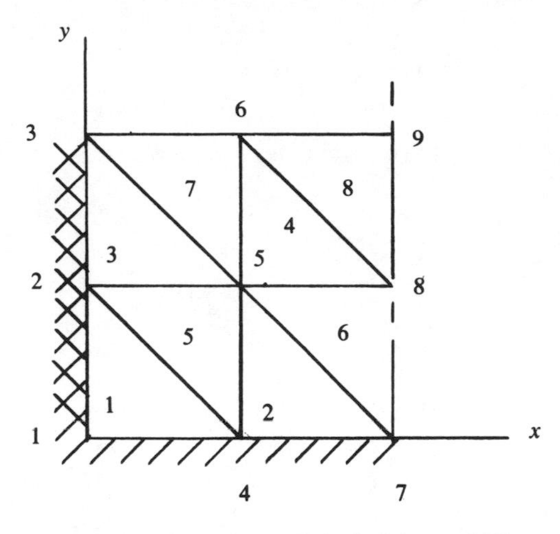

Figure 7-7. Element Subdivision of Plate

Table 7-2
Input Load Data for Hydrostatically Loaded Plate

Element No.	*Centroidal Pressure, p*	$\frac{\partial p}{\partial y}$	$\frac{\partial p}{\partial x}$
7, 8	0.350	−.350	0
3, 4	0.700	−.350	0
5, 6	1.400	−.350	0
1, 2	1.750	−.350	0

Table 7-3
Component Numbers of Known Displacements (all disp. = 0)

Node No.	w	$\frac{\partial w}{\partial x}$	$\frac{\partial w}{\partial y}$	$\frac{\partial^2 w}{\partial x^2}$	$\frac{\partial^2 w}{\partial x\,\partial y}$	$\frac{\partial^2 w}{\partial y^2}$
1	1	2	3	4	5	6
2	7	8	9		11	12
3	13	14	15		17	18
4	19	20		22		24
5						
6						36
7	37	38		40	41	42
8		44			47	
9		50			53	54

```
         MAIN PB11B  JUNE 25 1973
PLATE IN BENDING BY TRIANGULAR ELEMENTS (18 DEG. OF F.)

CASE TITLE  ---  GLASS PLATE UNDER HYDROSTATIC PRESSURE

YOUNGS MODULUS =  .100E+08   POISSONS RATIO = .200
PLATE THICKNESS = 1.250

NODE NO.    X-COORD     Y-COORD
    1      .0000E+00   .0000E+00
    2      .1000E-01   .3000E+02
    3      .2000E-01   .6000E+02
    4      .3000E+02   .0000E+00
    5      .3001E+02   .3000E+02
    6      .3002E+02   .6000E+02
    7      .6000E+02   .0000E+00
    8      .6001E+02   .3000E+02
    9      .6002E+02   .6000E+02

ELEM.NO.    CORNER NODE NUMBERS     PRESSURE     DP/DX        DP/DY
   1          1     4     2          1.750        .000        -.035
   2          4     7     5          1.750        .000        -.035
   3          2     5     3           .700        .000        -.035
   4          5     8     6           .700        .000        -.035
   5          2     4     5          1.400        .000        -.035
   6          5     7     8          1.400        .000        -.035
   7          3     5     6           .350        .000        -.035
   8          6     8     9           .350        .000        -.035

BAND WIDTH = 24

NUMBER OF DIFFERENT ELEMENT TYPES =   2

LOCAL ROTATED AXES
NODE NO.   THETA(DEG)

NODES HAVING MODIFIED CURVATURES
   6
   9

  KNOWN NON-ZERO LOADS
COMPONENT NUMBER   LOAD
      KNOWN DISPLACEMENTS
COMPONENT NUMBER   DISPLACEMENT
          1          .0000E+00            19          .0000E+00
          2          .0000E+00            20          .0000E+00
          3          .0000E+00            22          .0000E+00
          4          .0000E+00            24          .0000E+00
          5          .0000E+00            36          .0000E+00
          6          .0000E+00            37          .0000E+00
          7          .0000E+00            38          .0000E+00
          8          .0000E+00            40          .0000E+00
          9          .0000E+00            41          .0000E+00
         11          .0000E+00            42          .0000E+00
         12          .0000E+00            44          .0000E+00
         13          .0000E+00            47          .0000E+00
         14          .0000E+00            50          .0000E+00
         15          .0000E+00            53          .0000E+00
         17          .0000E+00            54          .0000E+00
         18          .0000E+00

NODE NO.      DISPLACEMENTS REFERRED TO GLOBAL AXES
   1    -.1738E-14  .2157E-15  .1940E-15 -.3675E-16  .8660E-16  .5885E-16
   2     .5848E-14  .8346E-15  .3617E-16  .3699E-03  .4229E-16  .5896E-16
   3     .6982E-14  .9869E-15 -.3364E-15  .4485E-03 -.1794E-15  .2042E-16
```

Figure 7-8. Output Processed by PB11B: Glass Plate

4	.6745E-15	.1713E-15	.3839E-02	.2500E-16	.1296E-03	-.4975E-16
5	.8435E-01	.3439E-02	.1451E-02	-.6795E-04	.7470E-04	-.7205E-04
6	.1102E+00	.4897E-02	.6962E-03	-.5971E-04	.5495E-04	.1194E-04
7	.2704E-14	.7774E-16	.6124E-02	.1211E-16	.8379E-16	-.4072E-16
8	.1418E+00	.5335E-15	.2772E-02	-.1319E-03	.1748E-16	-.1250E-03
9	.1955E+00	.5638E-15	.1640E-02	-.2159E-03	-.9228E-16	.4318E-04

NODE NO.	MXX	MYY	MXY	SXX	SYY	SXY
1	-.423E-10	.873E-10	.117E-09	.163E-09	-.335E-09	-.451E-09
2	.627E+03	.125E+03	.574E-10	-.241E+04	-.482E+03	-.220E-09
3	.760E+03	.152E+03	-.243E-09	-.292E+04	-.584E+03	.935E-09
4	.255E-10	-.759E-10	.176E+03	-.980E-10	.291E-09	-.675E+03
5	-.140E+03	-.145E+03	.101E+03	.536E+03	.558E+03	-.389E+03
6	-.972E+02	-.137E-10	.745E+02	.373E+03	.527E-10	-.286E+03
7	.672E-11	-.649E-10	.114E-09	-.258E-10	.249E-09	-.436E-09
8	-.266E+03	-.257E+03	.237E-10	.102E+04	.986E+03	-.910E-10
9	-.351E+03	-.118E-10	-.125E-09	.135E+04	.453E-10	.481E-09

THETA	PS1	PS2
-30.6	.429E-09	-.602E-09
.0	-.241E+04	-.482E+03
-.0	-.292E+04	-.584E+03
45.0	-.675E+03	.675E+03
44.2	.158E+03	.936E+03
-28.4	.528E+03	-.155E+03
36.3	-.346E-09	.569E-09
-.0	.102E+04	.986E+03
.0	.135E+04	.453E-10

Figure 7-8 (continued)

Solution. The sketch of the equilateral plate shown in Figure 7-9a, has one axis of symmetry and, hence, only one-half of the plate is analyzed. The subdivided plate is shown in Figure 7-9b. Because two of the plate edges are not parallel to the coordinate axes, a rotation of the axes system of 30° clockwise at nodes 5 and 7 is used as input data to the program.

The load, being uniform, is taken care of by specifying a load of 1000 pounds per square inch at the centroid of each element with no variation in either the x or the y directions.

The boundary conditions are given in Table 7-4 and the computer output for this example is shown in Figure 7-10.

Comments on Solutions

The three cases that have been solved above are the same as the examples used by Bell. The results agree with those of Bell and both Finite Element solutions agree quite well with the analytical solutions. The largest error occurred in Problem 2 where M_{xx} at node 3 has an error of 9.7 percent, the next largest is in Problem 1 where M_{xy} at node 1 is 1.4 percent in error. When the wide variation in moment over the region of the chosen elements is considered, it is remarkable that this accuracy is attained.

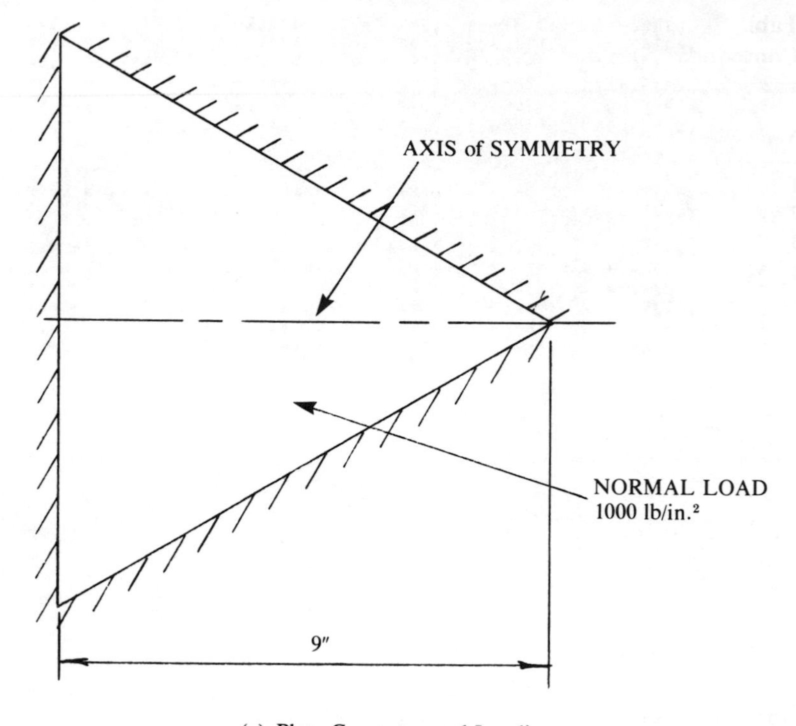

(a) Plate Geometry and Loading

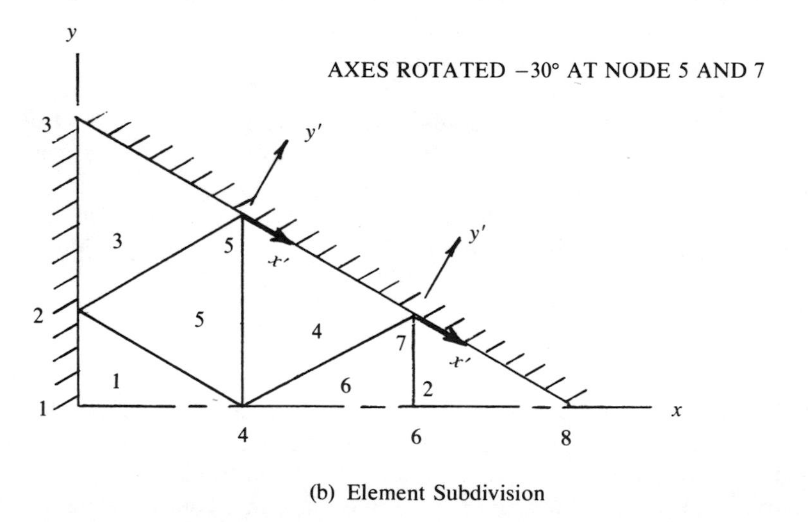

(b) Element Subdivision

Figure 7-9. Triangular Plate with Uniformly Distributed Load (a) Plate Geometry and Loading (b) Element Subdivision

Table 7-4
Component Numbers of Known Displacements (all disp. = 0)

Node No.	w	$\frac{\partial w}{\partial x}$	$\frac{\partial w}{\partial y}$	$\frac{\partial^2 w}{\partial x^2}$	$\frac{\partial^2 w}{\partial x\,\partial y}$	$\frac{\partial^2 w}{\partial y^2}$
1	1		3	4	5	6
2	7		9	10		12
3	13	14	15	16	17	18
4			21		23	
5	25	26		28		30
6			33		35	
7	37	38		40		42
8	43	44	45	46	47	48

```
          MAIN PB11B   JUNE 25 1973
 PLATE IN BENDING BY TRIANGULAR ELEMENTS (18 DEG. OF F.)

 CASE TITLE  ---  TRIANGULAR PLATE   UNIFORM LOAD   SIMPLY SUPPORTED

 YOUNGS MODULUS =   .300E+08    POISSONS RATIO = .300
 PLATE THICKNESS = 1.000

 NODE NO.     X-COORD       Y-COORD
      1     .0000E+00     .0000E+00
      2     .0000E+00     .1732E+01
      3     .0000E+00     .5196E+01
      4     .3000E+01     .0000E+00
      5     .3000E+01     .3464E+01
      6     .6000E+01     .0000E+00
      7     .6000E+01     .1732E+01
      8     .9000E+01     .0000E+00

 ELEM.NO.     CORNER NODE NUMBERS     PRESSURE       DP/DX          DP/DY
    1          1     4     2        1000.000         .000           .000
    2          6     8     7        1000.000         .000           .000
    3          2     5     3        1000.000         .000           .000
    4          4     7     5        1000.000         .000           .000
    5          2     4     5        1000.000         .000           .000
    6          4     6     7        1000.000         .000           .000

 BAND WIDTH = 24

 NUMBER OF DIFFERENT ELEMENT TYPES =     4

 LOCAL ROTATED AXES
NODE NO.     THETA(DEG)
   5           -30.0
   7           -30.0

 NODES HAVING MODIFIED CURVATURES

   KNOWN NON-ZERO LOADS
 COMPONENT NUMBER    LOAD
```

Figure 7-10. Output Processed by PB11B: Triangular Plate

KNOWN DISPLACEMENTS

COMPONENT NUMBER	DISPLACEMENT
1	.0000E+00
3	.0000E+00
4	.0000E+00
5	.0000E+00
6	.0000E+00
7	.0000E+00
9	.0000E+00
10	.0000E+00
12	.0000E+00
13	.0000E+00
14	.0000E+00
15	.0000E+00
16	.0000E+00
17	.0000E+00
18	.0000E+00
21	.0000E+00
23	.0000E+00
25	.0000E+00
26	.0000E+00
28	.0000E+00
30	.0000E+00
33	.0000E+00
35	.0000E+00
37	.0000E+00
38	.0000E+00
40	.0000E+00
42	.0000E+00
43	.0000E+00
44	.0000E+00
45	.0000E+00
46	.0000E+00
47	.0000E+00
48	.0000E+00

NODE NO.	DISPLACEMENTS REFERRED TO GLOBAL AXES					
1	.2988E-16	.1369E-02	-.4224E-16	-.4822E-16	-.2154E-15	.1350E-15
2	.2116E-16	.1090E-02	-.1508E-17	-.2493E-15	-.3101E-03	.4916E-16
3	-.1550E-15	-.8952E-17	.7680E-17	.1423E-15	.3327E-16	.4426E-17
4	.2457E-02	-.1224E-05	-.4251E-15	-.5447E-03	-.3587E-16	-.5462E-03
5	.1875E-15	-.5459E-03	-.9455E-03	-.2733E-03	-.1578E-03	.2733E-03
6	.9216E-03	-.6777E-03	-.1744E-15	.1369E-03	.8842E-16	-.6820E-03
7	.6628E-16	-.5466E-03	-.9468E-03	.2765E-03	.1597E-03	-.2765E-03
8	-.5164E-16	-.1464E-16	-.4185E-15	.3888E-16	.5439E-15	.6213E-16

Figure 7-10 (continued)

Experience shows that it is essential to have a systematic way of determining the known displacement components. The tabular arrangement shown in the examples is recommended, as it enables a comparison to be made to ensure that the nodes have been treated consistantly.

In some cases, ambiguities can not be avoided. To illustrate: in Problem 2, if node 3 is treated as a point on the free edge, the warping component, number 17, is an unknown.

In other cases, representation of the boundary conditions must remain imperfect. This is the case in Problem 3 at node 3 and at node 8 where skewed axes would be required for a perfect representation of the boundary conditions. The example shows that the error from this source is insignificant.

8 General Comments

Limits Imposed by Dimensions of Arrays

In all programs, the dimensions of the arrays impose certain restrictions on their use. The programs that have been presented were written primarily as teaching devices to be used for solving comparatively small problems. So that many programs could be constantly in the memory of the machine, and, hence, readily available, the arrays were made rather small. The limitations on any program can be found by examining the DIMENSION statements. For example, Line 4 of Figure 4-5 reveals that a system being treated by SP44B cannot have more than 260 degrees of freedom, that is, 130 nodes. From Line 4, it is also evident that the bandwidth cannot exceed 46. A check is made during execution to ensure that this limit is not exceeded through the value of LBAND which is established by the statement on Line 9.

Line 5 restricts the number of triangular elements to 90 while Line 6 restricts the number of bar elements to 40. The arrays in Line 7 and 8 have dimensions that do not vary with the case being executed.

Generally, in engineering practice the number of elements is made large to ensure good accuracy in the results. Hence, the DIMENSION statements would have to be altered to accommodate the larger number of nodes and elements, and larger bandwidth. The arrays have been segregated in the DIMENSION statements to facilitate such changes. In addition to the DIMENSION statements, only one other change is required; the LBAND = statement must be altered so that the value of LBAND matches the first value in the S array. It is essential that these values be equal.

Reliability of Data Cards

When solving problems with large numbers of elements, there is a high probability that an error will be made in the coordinates of a point or in the number of one of the corner nodes. The programs print all of this information but an error can still escape notice. Such an error results in a case being solved that differs from the intended case. Often the presence of an error is obvious from the printed stresses; but, if the problem is one in which the solution is not known in advance, the error caused by faulty data may go undetected.

The best practical method for ensuring a reliable data deck is to write a separate program that will read the data deck and make a plot of the elements. Gross errors in data are very conspicuous on such a figure.

Interpretation of Stress Values

The Finite Element Method is by its nature approximate; the approximation, in general, improving with finer subdivision of the stressed body. It is good practice to solve any problem several times using a finer subdivision each time, stopping when the stresses no longer change with finer subdivision.

The stresses may still require some interpretation, and a plot of stress variation across critical sections is often helpful. Wherever possible, the stresses on a section should be checked for equilibrium with the applied loads.

When very large numbers of elements are used, some important stress values may escape notice in the mass of printed values. It may then be worth while to alter the program so that stresses are punched as well as printed. The punched output along with the input data cards can then be processed by another program to plot the elastic body with the stresses plotted to scale directly on the figure. The best representation of stress is a plot of stress contours on the diagram of the elastic body, however the programming required for this is quite lengthy.

Precision and Round-Off Error

When solving for the unknown displacement components by the method of Payne and Irons and Gauss' elimination scheme, we are in effect inverting a stiffness matrix. Inverting a singular matrix is impossible and if the conditions that would produce a singular matrix exist in a problem, a division by zero will be encountered when attempting to carry out the elimination process. In stress problems, unless an error has been made, the matrix will never be singular; it can, however, approach this state, and then round-off errors will drastically reduce the number of significant figures in the output.

We have recommended that each problem be solved several times, making the elements smaller and more numerous with each repetition. With each step the numerical solution approaches more closely the theoretically correct solution. However, with each step the matrix comes closer to being singular and so round-off errors become more important. In extreme cases the round-off error can completely dominate so that there are no significant figures in the output. This imposes a limit on the fineness of the subdivision that can be used when the body is being broken into elements.

In most applications, if double precision is used the effect of round-off does not seriously influence the numerical values of stress. This topic is discussed by Rosanoff and Ginsberg [11], where a method is presented which determines the amount of round-off error. Unless a study is made to determine the effect of round-off, it is recommended that double precision be used.

Appendix: Subroutines

```
      SUBROUTINE AB01B(IE,NCON,XY,T,AB,VOL)
C   GENERATES A-BAR MATRIX AND VOLUME    W. H. BOWES   AUG 30 1971
      DIMENSION NCON(3,1),XY(2,1),AB(3,3),T(1)
      DO 4 I=1,3
      NC=NCON(I,IE)
      AB(I,1)=1.
      AB(I,2)=XY(1,NC)
    4 AB(I,3)=XY(2,NC)
      AREA=(AB(2,2)-AB(1,2))*(AB(3,3)-AB(1,3))-(AB(3,2)-AB(1,2))*(AB(2,3
     1)-AB(1,3))
      AREA=.5*AREA
      VOL=AREA*T(IE)
      IF(AREA)6,6,8
    6 VOL=-VOL
    8 RETURN
      END
```

```
      SUBROUTINE AI01B(AB,AI)
C   GENERATES A-INVERSE        W. H. BOWES   AUG 30 1971
      DIMENSION AB(3,3),AI(6,6)
      DO 2 I=1,3
      DO 2 J=1,3
      AI(I,J)=AB(I,J)
      AI(I+3,J)=0.
      AI(I,J+3)=0.
    2 AI(I+3,J+3)=AB(I,J)
      RETURN
      END
```

```
      SUBROUTINE AI02B(IE,NCON,XY,T,AI,X,Y)
C   GENERATES A-INVERSE AND LOCAL COORDS OF CORNERS OF LINEAR
C     STRAIN TRIANGLE       W H BOWES   SEPT 21 1971
      DIMENSION NCON(6,1),XY(2,1),T(1),AI(12,12),X(3),Y(3),AB(6,6)
      DIMENSION XX(6),YY(6)
      XC=0.
      YC=0.
      DO 4 I=1,3
      NC=NCON(2*I-1,IE)
      X(I)=XY(1,NC)
      Y(I)=XY(2,NC)
      XC=XC+X(I)/3.
    4 YC=YC+Y(I)/3.
      DO 6 I=1,3
      X(I)=X(I)-XC
      Y(I)=Y(I)-YC
      II=2*I-1
      XX(II)=X(I)
    6 YY(II)=Y(I)
      XX(2)=(XX(1)+XX(3))*.5
      XX(4)=(XX(3)+XX(5))*.5
      XX(6)=(XX(5)+XX(1))*.5
      YY(2)=(YY(1)+YY(3))*.5
      YY(4)=(YY(3)+YY(5))*.5
      YY(6)=(YY(5)+YY(1))*.5
      DO 7 I=2,6,2
      NCT=NCON(I,IE)
      XY(1,NCT)=XX(I)+XC
    7 XY(2,NCT)=YY(I)+YC
      DO 8 I=1,6
      AB(I,1)=1.
      AB(I,2)=XX(I)
      AB(I,3)=YY(I)
```

```
      AB(I,4)=XX(I)*XX(I)
      AB(I,5)=XX(I)*YY(I)
    8 AB(I,6)=YY(I)*YY(I)
      CALL IN12B(AB,6)
      DO 10 I=1,6
      DO 10 J=1,6
      AI(I,J)=AB(I,J)
      AI(I+6,J)=0.
      AI(I,J+6)=0.
   10 AI(I+6,J+6)=AB(I,J)
      RETURN
      END

      SUBROUTINE AM21B(XX,YY,A,CO,SI,FL,ND)
C  GENERATES A-MATRIX ETC FOR TRIANGULAR ELEMENT (18 DOF) IN
C        BENDING      W H BOWES    JAN 31 1972
      DIMENSION XX(4),YY(4),A(21,21),CO(3),SI(3),FL(3),F(21),ND(4)
      DO 40 III=1,2
      DO 40 II=1,3
      GO TO (1,2),III
    1 X=.5*(XX(II)+XX(II+1))
      Y=.5*(YY(II)+YY(II+1))
      IL=3
      GO TO 3
    2 X=XX(II)
      Y=YY(II)
      IL=6
    3 DO 30 I=1,IL
      IA=6*(II-1)+I
      J=0
      DO 30 K=1,6
      DO 30 L=1,K
      J=J+1
      IX=K-L
      IY=L-1
      IF(IX)4,4,6
    4 F(J)=1.
      IF(IY)10,10,5
    5 F(J)=Y**IY
      GO TO 10
    6 IF(IY)7,7,9
    7 F(J)=X**IX
      GO TO 10
    9 F(J)=X**IX*Y**IY
   10 GO TO (11,12,13,14,15,16),I
   11 C=1.
      IF=J
      GO TO 20
   12 C=IX
      IF=J-K+1
      GO TO 20
   13 C=IY
      IF=J-K
      GO TO 18
   14 C=IX*(IX-1)
      IF=J-2*K+3
      GO TO 20
   15 C=IX*IY
      IF=J-2*K+2
      GO TO 18
   16 C=IY*(IY-1)
      IF=J-2*K+1
   18 IF(IF)19,19,20
   19 FF=0.
      GO TO 21
   20 FF=F(IF)
   21 A(IA,J)=C*FF
   30 CONTINUE
      GO TO (32,40),III
   32 DX=XX(II+1)-XX(II)
      DY=YY(II+1)-YY(II)
```

```
      XY=SQRT(DX*DX+DY*DY)
      FL(II)=XY
      C=DX/XY
      S=DY/XY
      CO(II)=C
      SI(II)=S
      IL=II+18
      IR=6*II-4
      IRR=6*II-3
      DO 38 J=1,21
   38 A(IL,J)=(-S*A(IR,J)+C*A(IRR,J))*ND(II)
   40 CONTINUE
      RETURN
      END


      SUBROUTINE AS02B(E,S,K,LBAND)
C   ADDS  E  TO  S  FOR SHELL ELEMENT     W H BOWES     SEPT 21 1971
      DIMENSION S(LBAND,1),E(6,6)
      DO 2 J=1,6
      DO 2 I=J,6
      JJ=3*(K-1)+J
      II=I-J+1
    2 S(II,JJ)=S(II,JJ)+E(I,J)
      RETURN
      END


      SUBROUTINE AS03B(IE,NCON,E,FE,S,F,NF,NN,KE,LBAND,IT)
C   ADDS  E  AND  FE  TO  S  AND  F      FEB 10 1972
C        E=ELEMENT STIFFNESS
C        FE=FORCES AT NODES
C        NF=NUMBER OF DOF AT EACH NODE
C        NN=NUMBER OF NODES IN ELEMENT
C        KE=DEGREES OF FREEDOM IN ELEMENT (KE=NF*NN)
C        IT=1 WHEN ORDER OF DELTAS IS, ALL X, ALL Y ETC
C        IT=2 WHEN ORDER OF DELTAS IS, ALL AT NODE 1, ALL AT NODE 2 ETC
      DIMENSION NCON(NN,1),E(KE,KE),FE(1),S(LBAND,1),F(1)
      GO TO (1,2),IT
    1 K1=NN
      K2=1
      K3=NN
      GO TO 3
    2 K1=1
      K2=NF
      K3=NF
    3 DO 8 JA=1,NF
      DO 8 JB=1,NN
      J=K1*JA+K2*JB-K3
      JJ=NF*(NCON(JB,IE)-1)+JA
      F(JJ)=F(JJ)+FE(J)
      DO 8 IA=1,NF
      DO 8 IB=1,NN
      I=K1*IA+K2*IB-K3
      II=NF*(NCON(IB,IE)-1)+IA-JJ+1
      IF(II)8,8,6
    6 S(II,JJ)=S(II,JJ)+E(I,J)
    8 CONTINUE
      RETURN
      END


      SUBROUTINE AS06B(IE,NCON,E,S,NF,NN,KE,LBAND)
C   ADDS E TO S           W H BOWES  SEPT 10 1971
      DIMENSION NCON(NN,1),S(LBAND,1),E(KE,KE)
      DO 5 IA=1,NF
      DO 5 IB=1,NN
      DO 5 JA=1,NF
      DO 5 JB=1,NN
      JJ=NF*(NCON(JB,IE)-1)+JA
      II=NF*(NCON(IB,IE)-1)+IA-JJ+1
      IF(II)5,5,3
```

```
    3 I=NN*(IA-1)+IB
      J=NN*(JA-1)+JB
      S(II,JJ)=S(II,JJ)+E(I,J)
    5 CONTINUE
      RETURN
      END


      SUBROUTINE BC01B(B)
C   FILLS IN CONSTANT ELEMENTS OF B FOR CY, SHELLS  W H BOWES AUG 30 1971
      DIMENSION B(4,6)
      DO 2 I=1,4
      DO 2 J=1,6
    2 B(I,J)=0.
      B(1,2)=1.
      B(3,5)=2.
      RETURN
      END


      SUBROUTINE BV01B(B,X1,X2,Y1,Y2,FL,Z)
C   FILLS IN VARIABLE ELEMENTS OF B FOR CY, SHELLS  W H BOWES AUG 30 1971
      DIMENSION B(4,6)
      Y=Y1+Z*(Y2-Y1)
      IF(Y-.001)2,2,4
    2 Y=.001
    4 RY=1./Y
      XP=Z*FL
      SP=(Y2-Y1)/FL
      CP=(X2-X1)/FL
      B(2,1)=SP*RY
      B(4,4)=B(2,1)
      B(2,2)=XP*B(2,1)
      B(4,5)=2.*B(2,2)
      B(4,6)=1.5*XP*B(4,5)
      B(2,3)=CP*RY
      B(2,4)=XP*B(2,3)
      B(2,5)=XP*B(2,4)
      B(2,6)=XP*B(2,5)
      B(3,6)=6.*XP
      RETURN
      END

      SUBROUTINE CC01B(C,FL)
C   FILLS IN C-MATRIX FOR CY, SHELL ELEMENT  W H BOWES  AUG 30 1971
      DIMENSION C(6,6)
      DO 2 I=1,6
      DO 2 J=1,6
    2 C(I,J)=0.
      C(1,1)=1.
      C(2,3)=1.
      C(3,4)=1.
      C(4,1)=1.
      C(5,3)=1.
      C(6,4)=1.
      C(4,2)=FL
      C(5,4)=FL
      C(5,5)=FL**2
      C(5,6)=FL**3
      C(6,5)=2.*FL
      C(6,6)=3.*FL**2
      RETURN
      END


      SUBROUTINE CN02B(LBAND,NCON,T,DF,NE,NBAND)
C   READS CORNER NUMBERS,THICKNESSES AND TEMP, OF TRIANGULAR
C       ELEMENTS    W H BOWES   AUG 30 1971
      DIMENSION NCON(3,100),T(100),DF(100)
      WRITE(3,100)
      NE=0
      NBAND=0
```

```
      2 READ(2,102)I,N1,N2,N3,TH,TEMP
        IF(I)20,20,3
      3 NCON(1,I)=N1
        NCON(2,I)=N2
        NCON(3,I)=N3
        T(I)=1.
        DF(I)=TEMP
        IF(TH)4,5,4
      4 T(I)=TH
      5 ND=IABS(N1-N2)
        IF(NBAND-ND)6,7,7
      6 NBAND=ND
      7 ND=IABS(N2-N3)
        IF(NBAND-ND)8,9,9
      8 NBAND=ND
      9 ND=IABS(N3-N1)
        IF(NBAND-ND)10,11,11
     10 NBAND=ND
     11 WRITE(3,103)I,N1,N2,N3,T(I),TEMP
        NE=NE+1
        GO TO 2
     20 NBAND=2*NBAND+2
        WRITE(3,105)NBAND
        IF(LBAND-NBAND)22,24,24
     22 WRITE(3,106)
        CALL EXIT
     24 RETURN
    100 FORMAT('0ELEM.NO.     CONNECTING NODES NUMBERED  THICKNESS  TEMP')
    102 FORMAT(4I5,2F10.5)
    103 FORMAT(I6,4X,3I8,7X,F6.2,2X,F6.1)
    105 FORMAT('0BAND WIDTH =',I5)
    106 FORMAT('0BAND WIDTH IS TOO LARGE')
        END


        SUBROUTINE CN05B(LBAND,NCON,NE,NBAND)
C   READS CORNER NUMBERS OF QUAD ELEMENTS    W H BOWES  AUG 30 1971
        DIMENSION NCON(4,1),X(4),Y(4),K(4)
        NBAND=0
        NE=0
        WRITE(3,100)
      6 READ(2,102)N,K
        IF(N)20,20,8
      8 NE=NE+1
        DO 10 I=1,3
        II=I+1
        DO 10 J=II,4
        IF(IABS(K(I)-K(J))-NBAND)10,10,9
      9 NBAND=IABS(K(I)-K(J))
     10 CONTINUE
        DO 11 I=1,4
     11 NCON(I,N)=K(I)
        WRITE(3,104)N,K
        GO TO 6
     20 NBAND=2*(NBAND+1)
        IF(LBAND-NBAND)22,24,24
     22 WRITE(3,107)
        CALL EXIT
     24 WRITE(3,108)NBAND
        RETURN
    100 FORMAT('0ELEM.NO.     CORNER NODE NUMBERS')
    102 FORMAT(5I5)
    104 FORMAT(I6,5X,4I5)
    106 FORMAT(I3)
    107 FORMAT('0BAND WIDTH TOO LARGE, EXECUTION TERMINATED')
    108 FORMAT('0BAND WIDTH ='I3)
        END


        SUBROUTINE CN06B(LBAND,NN,NF,NCON,T,NE,NBAND)
C   READS ELEMENT DATA FO L S T     W H BOWES  SEPT 21 1971
        DIMENSION NCON(NN,1),T(1),N(12)
```

```
      WRITE(3,100)
      NE=0
      NBAND=0
      NN1=NN-1
    2 READ(2,102)I,TH,N
      IF(I)20,20,3
    3 DO 4 J=1,NN
    4 NCON(J,I)=N(J)
      T(I)=1.
      IF(TH)6,6,5
    5 T(I)=TH
    6 WRITE(3,103)I,T(I),(N(J),J=1,NN)
      DO 10 J=1,NN1
      JJ=J+1
      DO 10 I=JJ,NN
      ND=IABS(N(J)-N(I))
      IF(ND-NBAND)10,10,8
    8 NBAND=ND
   10 CONTINUE
      NE=NE+1
      GO TO 2
   20 NBAND=NF*(NBAND+1)
      WRITE(3,105)NBAND
      IF(LBAND-NBAND)22,24,24
   22 WRITE(3,106)
      CALL EXIT
   24 RETURN
  100 FORMAT('0ELEM.NO.  THICKNESS    CONNECTING NODE NUMBERS')
  102 FORMAT(I5,F10.5,12I5)
  103 FORMAT(I6,6X,F6.2,5X,12I4)
  105 FORMAT('0BAND WIDTH =',I5)
  106 FORMAT('0   ****  BAND WIDTH EXCEEDS LIMIT   ****')
      END


      SUBROUTINE CN08B(LBAND,NBAND,NE,NCON,PR)
C  READS ELEMENT DATA FOR 18 DOF TRIANGLE IN BENDING
C             JUNE 6 1972     W H BOWES
      DIMENSION NCON(3,1),PR(3,1),NC(4),PCXY(3)
      WRITE (3,100)
      NE=0
      NBAND=0
    2 READ(2,102) IE,(NC(I),I=1,3),PCXY
      IF(IE)20,20,6
    6 WRITE(3,104) IE,(NC(I),I=1,3),PCXY
      NE=NE+1
      DO 8 I=1,3
      NCON(I,IE)=NC(I)
    8 PR(I,IE)=PCXY(I)
      NC(4)=NC(1)
      DO 12 I=1,3
      ND=IABS(NC(I+1)-NC(I))
      IF(ND-NBAND)12,12,10
   10 NBAND=ND
   12 CONTINUE
      GO TO 2
   20 NBAND=(NBAND+1)*6
      IF(LBAND-NBAND)22,24,24
   22 WRITE(3,106)
      CALL EXIT
   24 WRITE(3,108)NBAND
      RETURN
  100 FORMAT('0ELEM.NO.   CORNER NODE NUMBERS   PRESSURE    DP/DX
     1 DP/DY')
  102 FORMAT(4I5,4F10.5)
  104 FORMAT(I5,3X,3I7,4F12.3)
  106 FORMAT('0BAND WIDTH OVER LIMIT  EXECUTION TERMINATED')
  108 FORMAT('0BAND WIDTH =',I3)
      END
      SUBROUTINE CN10B(NCB,ARB,NB,NBAND)
C  READS DATA FOR BAR ELEMENTS          W H BOWES   OCT 14 1971
      DIMENSION NCB(2,1),ARB(1)
```

```
      NB=0
      WRITE(3,102)
    2 READ(2,104)IB,AREA,NN1,NN2
      IF(IB)10,10,4
    4 WRITE(3,105)IB,AREA,NN1,NN2
      ARB(IB)=AREA
      NCB(1,IB)=NN1
      NCB(2,IB)=NN2
      NB=NB+1
      ND=2*(IABS(NN1-NN2)+1)
      IF(ND-NBAND)2,2,6
    6 NBAND=ND
      GO TO 2
   10 WRITE(3,110)NBAND
      RETURN
  102 FORMAT('0BAR NUMBER    X-SECT AREA    CONNECTS NODES NO.')
  104 FORMAT(I5,F10.5,2I5)
  105 FORMAT(I7,9X,F6.2,7X,2I5)
  110 FORMAT('0BAND WIDTH =',I5)
      END


      SUBROUTINE CP01B(D,B,X1,X2,Y1,Y2,FL,A,H,P)
C  CALCULATES P-MATRIX FOR SHELL ELEMENT      W H BOWES  AUG 30 1971
      DIMENSION B(4,6),D(4,4),DBY(4,6),P(6,6),A(6),H(6)
      PL=3.14159*FL
      DO 2 I=1,6
      DO 2 J=1,6
    2 P(I,J)=0.
      DO 30 IZ=1,6
      Z=A(IZ)
      X=.5*FL*(Z+1.)
      Y=Y1+X*(Y2-Y1)/FL
      B(2,1)=(Y2-Y1)/(FL*Y)
      B(2,2)=X*B(2,1)
      B(2,3)=(X2-X1)/(FL*Y)
      B(2,4)=X*B(2,3)
      B(2,5)=X*B(2,4)
      B(2,6)=X*B(2,5)
      B(3,6)=6.*X
      B(4,4)=B(2,1)
      B(4,5)=2.*X*B(4,4)
      B(4,6)=1.5*X*B(4,5)
      DO 12 I=1,4
      DO 12 J=1,6
      DIJ=0.
      DO 10 IJ=1,4
   10 DIJ=DIJ+D(I,IJ)*B(IJ,J)*Y
   12 DBY(I,J)=DIJ
      DO 22 I=1,6
      DO 22 J=1,6
      PIJ=0.
      DO 20 IJ=1,4
   20 PIJ=PIJ+B(IJ,I)*DBY(IJ,J)*H(IZ)
   22 P(I,J)=P(I,J)+PIJ
   30 CONTINUE
      DO 40 I=1,6
      DO 40 J=1,6
   40 P(I,J)=PL*P(I,J)
      RETURN
      END


      SUBROUTINE DB02B(YM,GNU,D,B)
C  D AND B MATRICES FOR TRIANGULAR ELEMENT      W. H. BOWES  AUG 30 1971
      DIMENSION D(3,3),B(3,6)
      D(1,1)=YM/(1.-GNU*GNU)
      D(1,2)=GNU*D(1,1)
      D(1,3)=0.
      D(2,1)=D(1,2)
      D(2,2)=D(1,1)
      D(2,3)=0.
```

```
      D(3,1)=0.
      D(3,2)=0.
      D(3,3)=.5*(1.-GNU)*D(1,1)
      DO 4 I=1,3
      DO 4 J=1,6
    4 B(I,J)=0.
      B(1,2)=1.
      B(2,6)=1.
      B(3,3)=1.
      B(3,5)=1.
      RETURN
      END


      SUBROUTINE DB03B (YM,GNU,D,B,DBD)
C FILLS IN D AND CONSTANTS IN B AND DBD FOR L S T  W H BOWES SEPT 3 1971
      DIMENSION D(3,3),B(3,12),DBD(12,12)
      D(1,1)=YM/(1.-GNU*GNU)
      D(1,2)=GNU*D(1,1)
      D(1,3)=0.
      D(2,1)=D(1,2)
      D(2,2)=D(1,1)
      D(2,3)=0.
      D(3,1)=0.
      D(3,2)=0.
      D(3,3)=.5*(1.-GNU)*D(1,1)
      DO 4 I=1,3
      DO 4 J=1,12
    4 B(I,J)=0.
      B(1,2)=1.
      B(2,9)=1.
      B(3,3)=1.
      B(3,8)=1.
      DO 6 I=1,12
      DO 6 J=1,12
    6 DBD(I,J)=0.
      DBD(2,2)=D(1,1)
      DBD(9,2)=D(1,2)
      DBD(2,9)=DBD(9,2)
      DBD(3,3)=D(3,3)
      DBD(8,3)=DBD(3,3)
      DBD(3,8)=DBD(8,3)
      DBD(8,8)=DBD(3,3)
      DBD(9,9)=DBD(2,2)
      RETURN
      END


      SUBROUTINE ES02B(P,C,AM,E)
C  STIFFNESS MATRIX FOR SHELL ELEMENT   W H BOWES AUG 30 1971
      DIMENSION P(6,6),C(6,6),AM(6,6),E(6,6),X(6,6)
      DO 3 I=1,6
      DO 3 J=1,6
      XIJ=0.
      DO 2 IJ=1,6
    2 XIJ=XIJ+C(I,IJ)*AM(IJ,J)
    3 X(I,J)=XIJ
      DO 5 I=1,6
      DO 5 J=1,6
      EIJ=0.
      DO 4 IJ=1,6
    4 EIJ=EIJ+P(I,IJ)*X(IJ,J)
    5 E(I,J)=EIJ
      DO 7 I=1,6
      DO 7 J=1,6
      XIJ=0.
      DO 6 IJ=1,6
    6 XIJ=XIJ+C(IJ,I)*E(IJ,J)
    7 X(I,J)=XIJ
      DO 9 I=1,6
      DO 9 J=1,6
      EIJ=0.
```

```
      DO 8 IJ=1,6
    8 EIJ=EIJ+AM(IJ,I)*X(IJ,J)
    9 E(I,J)=EIJ
      RETURN
      END


      SUBROUTINE ES03B(AI,B,D,VOL,ET,E,FTE)
C   STIFFNESS MATRIX AND THERMAL FORCE VECTOR FOR CONSTANT STRAIN
C       TRIANGULAR ELEMENT    W H BOWES  AUG 30 1971
      DIMENSION D(3,3)
      DIMENSION AI(6,6),B(3,6),E(6,6),P1(3,6),P2(3,6),FTE(6),DE(3)
      DO 3 I=1,3
      DO 3 J=1,6
      P1IJ=0.
      DO 2 IJ=1,6
    2 P1IJ=P1IJ+B(I,IJ)*AI(IJ,J)
    3 P1(I,J)=P1IJ
      DO 5 I=1,3
      DO 5 J=1,6
      P2IJ=0.
      DO 4 IJ=1,3
    4 P2IJ=P2IJ+D(I,IJ)*P1(IJ,J)
    5 P2(I,J)=P2IJ
      DO 8 I=1,6
      DO 8 J=1,6
      EIJ=0.
      DO 6 IJ=1,3
    6 EIJ=EIJ+P1(IJ,I)*P2(IJ,J)
    8 E(I,J)=VOL*EIJ
      DO 10 I=1,3
   10 DE(I)=ET*(D(I,1)+D(I,2))
      DO 12 I=1,6
      FTE(I)=0.
      DO 12 IJ=1,3
   12 FTE(I)=FTE(I)+P1(IJ,I)*DE(IJ)*VOL
      RETURN
      END


      SUBROUTINE ES06B(V,AI,BDB,X,Y,E)
C   ELEMENT STIFFNESS FOR LINEAR STRAIN TRIANGLE     W H BOWES  SEPT 3 1971
      DIMENSION AI(12,12),BDB(12,12),P(12,12),X(3),Y(3),E(12,12)
      XX=0.
      YY=0.
      XY=0.
      DO 2 I=1,3
      XX=XX+X(I)*X(I)
      YY=YY+Y(I)*Y(I)
    2 XY=XY+X(I)*Y(I)
      XX=XX/12.
      YY=YY/12.
      XY=XY/12.
      BDB(4,4)=4.*XX*BDB(2,2)
      BDB(5,4)=2.*XY*BDB(2,2)
      BDB(4,5)=BDB(5,4)
      BDB(11,4)=2.*XX*BDB(9,2)
      BDB(4,11)=BDB(11,4)
      BDB(12,4)=4.*XY*BDB(9,2)
      BDB(4,12)=BDB(12,4)
      BDB(5,5)=YY*BDB(2,2)+XX*BDB(3,3)
      BDB(6,5)=2.*XY*BDB(3,3)
      BDB(5,6)=BDB(6,5)
      BDB(10,5)=2.*XX*BDB(3,3)
      BDB(5,10)=BDB(10,5)
      BDB(11,5)=XY*(BDB(9,2)+BDB(3,3))
      BDB(5,11)=BDB(11,5)
      BDB(12,5)=2.*YY*BDB(9,2)
      BDB(5,12)=BDB(12,5)
      BDB(6,6)=4.*YY*BDB(3,3)
      BDB(10,6)=4.*XY*BDB(3,3)
      BDB(6,10)=BDB(10,6)
```

```
      BDB(11,6)=.5*BDB(6,6)
      BDB(6,11)=BDB(11,6)
      BDB(10,10)=2.*BDB(10,5)
      BDB(11,10)=BDB(6,5)
      BDB(10,11)=BDB(11,10)
      BDB(11,11)=XX*BDB(2,2)+YY*BDB(3,3)
      BDB(12,11)=BDB(5,4)
      BDB(11,12)=BDB(12,11)
      BDB(12,12)=4.*YY*BDB(2,2)
      DO 6 I=1,12
      DO 6 J=1,12
      P(I,J)=0.
      DO 6 IJ=1,12
    6 P(I,J)=P(I,J)+BDB(I,IJ)*AI(IJ,J)
      DO 10 I=1,12
      DO 10 J=1,12
      E(I,J)=0.
      DO 8 IJ=1,12
    8 E(I,J)=E(I,J)+AI(IJ,I)*P(IJ,J)
   10 E(I,J)=V*E(I,J)
      RETURN
      END


      SUBROUTINE ES10B(IB,NCB,XY,ARB,YMB,S,LBAND)
C   CALCS BAR STIFFNESS AND ADDS TO S          W H BOWES   OCT 14 1971
      DIMENSION NCB(2,1),XY(2,1),ARB(1),S(LBAND,1)
      DIMENSION BK(4,4),XXYY(4)
      NN1=NCB(1,IB)
      NN2=NCB(2,IB)
      X=XY(1,NN2)-XY(1,NN1)
      Y=XY(2,NN2)-XY(2,NN1)
      AEL=ARB(IB)*YMB/(SQRT(X**2+Y**2))**3
      XXYY(1)=-X
      XXYY(2)=X
      XXYY(3)=-Y
      XXYY(4)=Y
      DO 6 I=1,4
      DO 6 J=1,4
    6 BK(I,J)=AEL*XXYY(I)*XXYY(J)
      DO 10 I1=1,2
      DO 10 I2=1,2
      DO 10 J1=1,2
      DO 10 J2=1,2
      JJ=2*NCB(J2,IB)-2+J1
      II=2*NCB(I2,IB)-1+I1-JJ
      IF(II)10,10,8
    8 I=2*(I1-1)+I2
      J=2*(J1-1)+J2
      S(II,JJ)=S(II,JJ)+BK(I,J)
   10 CONTINUE
      RETURN
      END


      SUBROUTINE ES18B(CQ,D,G,P,E)
C   CALCULATES STIFFNESS MATRIX FOR 18 DOF TRIANGLE IN BENDING
C             W H BOWES     FEB15 1972
      DIMENSION G(21,18),D(3,3),P(7,7),E(18,18),CQ(21,3)
      DIMENSION CI(21),CJ(21),Q(21,21)
      DO 8 I=1,21
      DO 8 J=1,21
    8 Q(I,J)=0.
      DO 32 KT=1,5
      GO TO (1,2,3,4,5),KT
    1 CV=4.*D(1,1)
      LII=0
      LJJ=0
      JCI=1
      JCJ=1
      GO TO 20
```

```
    2 CV=4.*D(1,2)
      LII=-2
      JCI=2
      GO TO 20
    3 CV=4.*D(1,1)
      LJJ=-2
      JCJ=2
      GO TO 20
    4 CV=4.*D(1,2)
      LII=0
      JCI=1
      GO TO 20
    5 CV=4.*D(3,3)
      LII=-1
      LJJ=-1
      JCI=3
      JCJ=3
   20 CONTINUE
      DO 30 KI=1,4
      III=(KI*(KI+3))/2+3
      DO 30 KJ=KI,4
      JJJ=(KJ*(KJ+3))/2+3
      DO 30 LI=1,KI
      ICQ=III+LI
      I=ICQ+LII
      DO 30 LJ=1,KJ
      JCQ=JJJ+LJ
      J=JCQ+LJJ
      II=KI+KJ+1-LI-LJ
      JJ=LI+LJ-1
   30 Q(I,J)=Q(I,J)+CV*CQ(ICQ,JCI)*CQ(JCQ,JCJ)*P(II,JJ)
   32 CONTINUE
      DO 34 I=1,21
      DO 34 J=I,21
   34 Q(J,I)=Q(I,J)
      DO 40 J=1,18
      DO 38 IC=1,21
      C=0.
      DO 36 IJ=1,21
   36 C=C+Q(IC,IJ)*G(IJ,J)
   38 CJ(IC)=C
      DO 40 I=1,18
      C=0.
      DO 39 IJ=1,21
   39 C=C+G(IJ,I)*CJ(IJ)
   40 E(I,J)=C
      RETURN
      END


      SUBROUTINE ESS1B(XY,NCON,KE,D,A,H,E)
C  STIFFNESS OF QUAD RING ELEMENT    W H BOWES  AUG 30 1971
      DIMENSION XY(2,1),NCON(4,1),D(4,4),A(3),H(3),E(8,8)
      DIMENSION XYE(4,2),B(4,8),FJ(2,2),EX(2,4),PX(4),PE(4),P(4),DB(4,8)
      DO 4 I=1,4
      IJ=NCON(I,KE)
      XYE(I,1)=XY(1,IJ)
    4 XYE(I,2)=XY(2,IJ)
      DO 7 J=1,8
      DO 6 I=1,4
    6 B(I,J)=0.
      DO 7 I=1,8
    7 E(I,J)=0.
      DO 40 IXH=1,3,2
      DO 40 IEH=1,3,2
      DO 40 IX=1,3
      DO 40 IE=1,3
      XI=(IXH-2)*A(IX)
      ETA=(IEH-2)*A(IE)
      PX(1)=1.-XI
      PX(2)=1.+XI
      PX(3)=PX(2)
```

```
      PX(4)=PX(1)
      PE(1)=1.-ETA
      PE(2)=PE(1)
      PE(3)=1.+ETA
      PE(4)=PE(3)
      Y=0.
      DO 10 I=1,4
      P(I)=PX(I)*PE(I)
   10 Y=Y+.25*P(I)*XYE(I,2)
      RY=1./Y
      EX(1,1)=-PE(1)
      EX(1,2)= PE(2)
      EX(1,3)= PE(3)
      EX(1,4)=-PE(4)
      EX(2,1)=-PX(1)
      EX(2,2)=-PX(2)
      EX(2,3)= PX(3)
      EX(2,4)= PX(4)
      DO 13 I=1,2
      DO 13 J=1,2
      FJIJ=0.
      DO 12 IJ=1,4
   12 FJIJ=FJIJ+.25*EX(I,IJ)*XYE(IJ,J)
   13 FJ(I,J)=FJIJ
      DET=FJ(1,1)*FJ(2,2)-FJ(2,1)*FJ(1,2)
      RDET=1./DET
      SJ=FJ(1,1)
      FJ(1,1)=FJ(2,2)*RDET
      FJ(2,2)=SJ*RDET
      FJ(2,1)=-FJ(2,1)*RDET
      FJ(1,2)=-FJ(1,2)*RDET
      DO 20 J=1,4
      JJ=J+4
      B(3,JJ)=.25*(RY*P(J))
      B(1,J)=.25*(FJ(1,1)*EX(1,J)+FJ(1,2)*EX(2,J))
      B(4,J)=.25*(FJ(2,1)*EX(1,J)+FJ(2,2)*EX(2,J))
      B(2,JJ)=.25*(FJ(2,1)*EX(1,J)+FJ(2,2)*EX(2,J))
   20 B(4,JJ)=.25*(FJ(1,1)*EX(1,J)+FJ(1,2)*EX(2,J))
      DO 25 I=1,4
      DO 25 J=1,8
      DBIJ=0.
      DO 24 IJ=1,4
   24 DBIJ=DBIJ+D(I,IJ)*B(IJ,J)
   25 DB(I,J)=DBIJ
      DO 30 I=1,8
      DO 30 J=1,8
      SIJ=0.
      DO 26 IJ=1,4
   26 SIJ=SIJ+B(IJ,I)*DB(IJ,J)
   30 E(I,J)=E(I,J)+6.28318*Y*DET*H(IX)*H(IE)*SIJ
   40 CONTINUE
      RETURN
      END


      SUBROUTINE EV02B (A,CO,SI,FL,ND,G)
C     ELIMINATES VARIABLES  (18 DOF TRI IN BENDING)  W H BOWES  JUNE 6 1973
      DIMENSION A(21,21),CO(3),SI(3),FL(3),G(21,18),H(3,18),ND(4)
      DIMENSION D(3,3)
      DO 2 I=1,3
      DO 2 J=1,18
    2 H(I,J)=0.
      DO 10 I=1,3
      S=SI(I)
      C=CO(I)
      F=FL(I)
      SIGN=ND(I)
      IJ=6*(I-1)
      IJJ=IJ+6
      IF(IJJ-18)5,4,4
    4 IJJ=IJJ-18
    5 H(I,IJ+2)=-.5*S*SIGN
```

```
      H(I,IJJ+2)=H(I,IJ+2)
      H(I,IJ+3)=.5*C*SIGN
      H(I,IJJ+3)=H(I,IJ+3)
      H(I,IJ+4)=-.125*F*S*C*SIGN
      H(I,IJJ+4)=-H(I,IJ+4)
      H(I,IJ+5)= .125*F*(C*C-S*S)*SIGN
      H(I,IJJ+5)=-H(I,IJ+5)
      H(I,IJ+6)=-H(I,IJ+4)
   10 H(I,IJJ+6)=-H(I,IJ+6)
      DO 20 I=1,21
      DO 20 J=1,18
      T=A(I,J)
      DO 18 IJ=1,3
   18 T=T+A(I,IJ+18)*H(IJ,J)
   20 G(I,J)=T
      RETURN
      END


      SUBROUTINE FP01B(PR,Y1,Y2,C,FL,AM,KE,F)
C   CONSISTENT LOAD VECTOR FOR PRESSURE IN SHELL   W H BOWES AUG 30 1971
      DIMENSION C(6,6),AM(6,6),F(1),FP(6),CFP(6)
      FP(3)=PR*.5*(Y1+Y2)*FL
      FP(4)=PR/6.*(Y1+2.*Y2)*FL**2
      FP(5)=PR/12.*(Y1+3.*Y2)*FL**3
      FP(6)=PR*0.05*(Y1+4.*Y2)*FL**4
      DO 4 I=1,6
      CFP(I)=0.
      DO 4 J=3,6
    4 CFP(I)=CFP(I)+C(J,I)*FP(J)
      DO 8 I=1,6
      FP(I)=0.
        DO 8 J=1,6
    8 FP(I)=FP(I)+AM(J,I)*CFP(J)
      DO 10 J=1,6
      JJ=3*KE-3+J
   10 F(JJ)=F(JJ)+FP(J)
      RETURN
      END


      SUBROUTINE FU10B(S,U,NBAND,NDF,NCN,LBAND)
C   CALCULATES F FROM S AND U. PRINTS F AND U.   W. H. BOWES AUG 30 1971
      DIMENSION S(LBAND,1),U(1),FC(6),UC(6)
      WRITE (3,100)
      NN=NDF/NCN
      DO 20 IN=1,NN
      DO 16 II=1,NCN
      IJ=(IN-1)*NCN+II
      FORCE=S(1,IJ)*U(IJ)
      DO 12 I=2,NBAND
      JL=IJ+1-I
      IF(JL)8,8,5
    5 FORCE=FORCE+S(I,JL)*U(JL)
    8 JR=IJ-1+I
      IF(NDF-JR)12,10,10
   10 FORCE=FORCE+S(I,IJ)*U(JR)
   12 CONTINUE
      UC(II)=U(IJ)
   16 FC(II)=FORCE
   20 WRITE(3,102) IN,(FC(II),II=1,NCN),(UC(II),II=1,NCN)
      RETURN
  100 FORMAT('ONODE NO.      FORCE AND DISPLACEMENT COMPONENTS')
  102 FORMAT(2X,I5,3X,10E11.4,/,98X,2E11.4)
      END


      SUBROUTINE GD02B(E,P,T,A,H,D)
C   GAUSS CONSTANTS AND D-MATRIX FOR SHELL ELEMENTS W H BOWES AUG 30 1971
      DIMENSION A(6),H(6),D(4,4)
      A(1)=-.93247
      A(2)=-.66121
```

```
      A(3)=-.23862
      A(4)=-A(3)
      A(5)=-A(2)
      A(6)=-A(1)
      H(1)=.17132
      H(2)=.36076
      H(3)=.46791
      H(4)=H(3)
      H(5)=H(2)
      H(6)=H(1)
      DO 5 I=1,4
      DO 5 J=1,4
    5 D(I,J)=0.
      D(1,1)=E*T/(1.-P*P)
      D(2,2)=D(1,1)
      D(1,2)=P*D(1,1)
      D(2,1)=D(1,2)
      D(3,3)=T*T/12.*D(1,1)
      D(4,4)=D(3,3)
      D(3,4)=P*D(3,3)
      D(4,3)=D(3,4)
      WRITE(3,100)E,P,T
      RETURN
  100 FORMAT('0YOUNGS MODULUS     =',E10.2,/,' POISSONS RATIO     =',F6.3,
     1/,' THICKNESS          =',F6.3)
      END


      SUBROUTINE GD03B(YM,GNU,A,H,D)
C   GENERATES GAUSS CONSTANTS AND D-MATRIX FOR RING ELEMENT
C         W H BOWES   AUG 30 1971
      DIMENSION A(3),H(3),D(4,4)
      A(1)=.23861919
      H(1)=.46791393
      A(2)=.66120939
      H(2)=.36076157
      A(3)=.93246951
      H(3)=.17132449
      DO 4 I=1,4
      DO 4 J=1,4
    4 D(I,J)=0.
      D(1,1)=YM*(1.-GNU)/((1.+GNU)*(1.-2.*GNU))
      D(2,2)=D(1,1)
      D(3,3)=D(1,1)
      D(4,4)=.5*YM/(1.+GNU)
      D(1,2)=D(1,1)*GNU/(1.-GNU)
      D(1,3)=D(1,2)
      D(2,1)=D(1,2)
      D(2,3)=D(1,2)
      D(3,1)=D(1,2)
      D(3,2)=D(1,2)
      RETURN
      END


      SUBROUTINE GE02B(F,U,S,IFX,NBAND,NDF,LBAND)
C   SOLVES FOR UNKNOWN U BY METHOD OF P AND I, GAUSSIAN ELIMINATION
C     AND BACK SUBSTITUTION         W H BOWES    FEB 22 1972
      DIMENSION S(LBAND,1),F(1),U(1),IFX(1)
      CP=10.**12
      DO 6 I=1,NDF
      IF(IFX(I))6,6,4
    4 S(1,I)=CP*S(1,I)
      F(I)=S(1,I)*U(I)
    6 CONTINUE
      IJ=1
      NA=NBAND
      NR=NDF-NBAND
   10 R=1./S(1,IJ)
      F(IJ)=R*F(IJ)
      IF(NA-1)30,30,12
   12 DO 20 II=2,NA
```

```
      C=S(II,IJ)
      S(II,IJ)=R*S(II,IJ)
      DO 18 JJ=2,II
      I=II+1-JJ
      J=IJ-1+JJ
   18 S(I,J)=S(I,J)-C*S(JJ,IJ)
      IIF=IJ-1+II
   20 F(IIF)=F(IIF)-C*F(IJ)
      NR=NR-1
      IF(NR)22,24,24
   22 NA=NA-1
   24 IJ=IJ+1
      GO TO 10
   30 NA=1
      IJ=NDF
      U(IJ)=F(IJ)
   32 IJ=IJ-1
      IF(IJ)34,34,36
   34 RETURN
   36 IF(NA-NBAND)38,40,40
   38 NA=NA+1
   40 C=F(IJ)
      DO 44 I=2,NA
      J=IJ-1+I
   44 C=C-S(I,IJ)*U(J)
      U(IJ)=C
      GO TO 32
      END


      SUBROUTINE GL01B(X1,X2,Y1,Y2,AM,FL)
C     GENERATES LAMBDA-MATRIX FOR SHELL ELEMENT       W H BOWES   AUG 30 1971
      DIMENSION AM(6,6)
      DO 2 I=1,6
      DO 2 J=1,6
    2 AM(I,J)=0.
      FL=SQRT((X2-X1)**2+(Y2-Y1)**2)
      AM(1,1)=(X2-X1)/FL
      AM(1,2)=(Y2-Y1)/FL
      AM(2,1)=-AM(1,2)
      AM(2,2)=AM(1,1)
      AM(3,3)=1.
      DO 4 I=1,3
      DO 4 J=1,3
    4 AM(I+3,J+3)=AM(I,J)
      RETURN
      END


      SUBROUTINE GN02B(S,U,F,IFX,NB,LBAND,TRIG)
C     ALTERS STIFFNESS ARRAY FOR GUIDED NODES      W H BOWES    AUG 30 1971
      DIMENSION S(LBAND,1),U(1),F(1),IFX(1),TRIG(1)
      WRITE(3,100)
    2 READ(2,102)NN,ALP,UK,FT
      IF(NN)4,4,6
    4 RETURN
    6 NN2=2*NN
      NN21=NN2-1
      IFX(NN21)=1
      U(NN21)=UK
      WRITE(3,106)NN,ALP,UK,FT
      S11=S(1,NN21)
      S21=S(2,NN21)
      S22=S(1,NN2)
      ALP=3.14159/180.*ALP
      C=COS(ALP)
      TRIG(NN21)=C
      C2=C*C
      SN=SIN(ALP)
      TRIG(NN2)=SN
      S2=SN*SN
      CS=C*SN
```

```
      FS=F(NN21)
      F(NN21)=C*F(NN21)+SN*F(NN2)
      F(NN2)=-SN*FS+C*F(NN2)+FT
      S(1,NN21)=C2*S11+2.*CS*S21+S2*S22
      S(2,NN21)=-CS*S11+(C2-S2)*S21+CS*S22
      S(1,NN2)=S2*S11-2.*CS*S21+C2*S22
      NB1=NB-1
      DO 30 I=2,NB1
      I1=I+1
      J=NN2-I
      IF(J)20,20,10
   10 S1=S(I,J)
      S2=S(I1,J)
      S(I,J)=C*S1+SN*S2
      S(I1,J)=-SN*S1+C*S2
   20 S1=S(I1,NN21)
      S2=S(I,NN2)
      S(I1,NN21)=C*S1+SN*S2
   30 S(I,NN2)=-SN*S1+C*S2
      GO TO 2
  100 FORMAT('0NODE NO.  ALPHA(DEG)  KNOWN U     TANG.FORCE')
  102 FORMAT(I5,3F10.5)
  106 FORMAT(I6,F11.1,4X,E10.3,2X,E10.3)
      END

      SUBROUTINE IN12B(A,N)
C   INVERTS AN N*N MATRIX, A          W H BOWES  SEPT 16 1971
      DIMENSION A(N,N),B(21,21)
      FN=N
      SUM=0.
      DO 4 I=1,N
      DO 2 J=1,N
    2 B(I,J)=0.
      SUM=SUM+ABS(A(I,I))
    4 B(I,I)=1.
      AVA=SUM/FN
      RER=.000001*AVA
      DO 40 I=1,N
      IF(ABS(A(I,I))-RER)6,6,20
    6 IF(I-N)10,7,10
    7 WRITE(3,100)
      CALL EXIT
   10 DO 12 II=I,N
      IF(ABS(A(II,I))-RER)12,12,14
   12 CONTINUE
      GO TO 7
   14 DO 16 JJ=1,N
      A(I,JJ)=A(I,JJ)+A(II,JJ)
   16 B(I,JJ)=B(I,JJ)+B(II,JJ)
   20 R=1./A(I,I)
      DO 22 J=1,N
      A(I,J)=R*A(I,J)
   22 B(I,J)=R*B(I,J)
      DO 37 II=1,N
      IF(II-I)32,37,32
   32 C=-A(II,I)
      DO 35 J=1,N
      A(II,J)=A(II,J)+C*A(I,J)
      B(II,J)=B(II,J)+C*B(I,J)
   35 CONTINUE
   37 CONTINUE
   40 CONTINUE
      DO 42 I=1,N
      DO 42 J=1,N
   42 A(I,J)=B(I,J)
      RETURN
  100 FORMAT('0      *****   MATRIX IS SINGULAR   *****')
      END

      SUBROUTINE KF01B(F,NDF)
```

```
C READS KNOWN FORCE COMPONENTS    W. H. BOWES    AUG 30 1971
      DIMENSION F(120)
      WRITE(3,100)
      DO 2 I=1,NDF
    2 F(I)=0.
    4 READ(2,104)I,FT
      IF(I)10,10,5
    5 F(I)=FT
      WRITE(3,106)I,FT
      GO TO 4
   10 RETURN
  100 FORMAT('0  KNOWN NON-ZERO LOADS',/,' COMPONENT NUMBER   LOAD')
  104 FORMAT(I5,F10.5)
  106 FORMAT(5X,I5,7X,E11.4)
      END


      SUBROUTINE KF02B(F)
C READS KNOWN FORCE COMPONENTS    W. H. BOWES  AUG 30 1971
      DIMENSION F(1)
      WRITE(3,100)
    4 READ(2,104)I,FT
      IF(I)10,10,5
    5 F(I)=F(I)+FT
      WRITE(3,106)I,FT
      GO TO 4
   10 RETURN
  100 FORMAT('0  KNOWN NON-ZERO LOADS',/,' COMPONENT NUMBER   LOAD')
  104 FORMAT(I5,F10.5)
  106 FORMAT(5X,I5,7X,E11.4)
      END


      SUBROUTINE KU01B(U,IHLD,NDF)
C  READS KNOWN DISPLACEMENT COMPONENTS    W. H. BOWES  AUG 30 1971
      DIMENSION U(120),IHLD(120)
      WRITE(3,100)
      DO 2 I=1,NDF
      U(I)=0.
    2 IHLD(I)=0
    4 READ(2,104)I,UT
      IF(I)10,10,5
    5 IHLD(I)=1
      U(I)=UT
      WRITE(3,106)I,UT
      GO TO 4
   10 RETURN
  100 FORMAT('0      KNOWN DISPLACEMENTS',/,' COMPONENT NUMBER   DISPLAC
     1EMENT')
  104 FORMAT(I5,F10.5)
  106 FORMAT(7X,I5,8X,E11.4)
      END


      SUBROUTINE LR01B(S,LBAND,NBAND,NU,T,NF)
C  TRANSFORMS  S  ( S = T-T * S * T)          W H BOWES   APRIL 30 1973
C  T IS TRANSFORMATION MATRIX
C  NF IS NO OF DOF AT A NODE (MUST NOT EXCEED 6)
C  NU IS DIMENSION OF LEADING UNIT MATRIX
      DIMENSION S(LBAND,1),T(NF,NF),P(6),TJ(6),FK(6,6)
      NBF=NBAND-NF+1
      DO 20 J=1,NF
      DO 4 I=1,NF
    4 TJ(I)=T(I,J)
      DO 12 I=1,NF
      SUM=0.
      DO 10 IJ=1,NF
      JJ=NU+IJ
      II=I+1-IJ
      IF(II)8,8,10
    8 JJ=JJ+II-1
      II=2-II
```

```
   10 SUM=SUM+S(II,JJ)*TJ(IJ)
   12 P(I)=SUM
      DO 20 IK=1,NF
      SUM=0.
      DO 14 IJ=1,NF
   14 SUM=SUM+T(IJ,IK)*P(IJ)
   20 FK(IK,J)=SUM
      DO 30 J=1,NF
      JJ=NU+J
      DO 30 I=J,NF
      II=I+1-J
   30 S(II,JJ)=FK(I,J)
      DO 40 IS=2,NBF
      JJ=NU+2-IS
      IF(JJ) 41,41,31
   31 DO 36 IP=1,NF
      SUM=0.
      DO 32 IJ=1,NF
      II=IS-1+IJ
   32 SUM=SUM+T(IJ,IP)*S(II,JJ)
   36 P(IP)=SUM
      DO 40 IJ=1,NF
      II=IS-1+IJ
   40 S(II,JJ)=P(IJ)
   41 DO 50 IS=2,NBF
      DO 46 IP=1,NF
      SUM=0.
      DO 42 IJ=1,NF
      JJ=NU+IJ
      II=IS+NF-IJ
   42 SUM=SUM+S(II,JJ)*T(IJ,IP)
   46 P(IP)=SUM
      DO 50 IJ=1,NF
      JJ=NU+IJ
      II=IS+NF-IJ
   50 S(II,JJ)=P(IJ)
      RETURN
      END


      SUBROUTINE LR02B(F,NU,T,NF)
C ALTERS F COMPONENTS FOR LOCAL AXES ROTATION   W H BOWES  JUNE 19 1973
      DIMENSION F(1),T(NF,NF),FF(10)
      DO 2 I=1,NF
    2 FF(I)=F(NU+I)
      DO 6 I=1,NF
      SUM=0.
      DO 4 IJ=1,NF
    4 SUM=SUM+T(IJ,I)*FF(IJ)
    6 F(NU+I)=SUM
      RETURN
      END


      SUBROUTINE RC02B(NF,XY,NDF)
C   READS COORDINATES OF NODES    W. H. BOWES    AUG 30 1971
C   NF IS NUMBER OF DEGREES OF FREEDOM AT EACH NODE
      DIMENSION XY(2,1)
      NDF=0
      WRITE(3,100)
    2 READ(2,102)I,X,Y
      IF(I)20,20,4
    4 XY(1,I)=X
      XY(2,I)=Y
      NDF=NDF+NF
      WRITE(3,104)I,X,Y
      GO TO 2
   20 RETURN
  100 FORMAT('0NODE NO.    X-COORD      Y-COORD')
  102 FORMAT(I5,2F10.5)
  104 FORMAT(I6,2X,2E12.4)
      END
```

```
      SUBROUTINE RC03B(XY,NE,NDF)
C READS COORDINATES OF NODES FOR CYLINDRICAL SHELL W H BOWES AUG 30 1971
      DIMENSION XY(2,60)
      NE=-1
      WRITE(3,100)
    2 READ(2,102)I,X,Y
      IF(I)20,20,4
    4 XY(1,I)=X
      XY(2,I)=Y
      NE=NE+1
      WRITE(3,104)I,X,Y
      GO TO 2
   20 NDF=3*NE+3
      RETURN
  100 FORMAT('0NODE NO.   X-COORD      Y-COORD')
  102 FORMAT(I5,2F10.5)
  104 FORMAT(I6,2X,2E12.4)
      END


      SUBROUTINE RC06B(NF,XY,NDF)
C  READS COORDINATES OF NODES    W. H. BOWES    SEPT 11 1971
      DIMENSION XY(2,1)
      WRITE(3,100)
      READ(2,102)NN
      NDF=NF*NN
    2 READ(2,102)I,X,Y
      IF(I)20,20,4
    4 XY(1,I)=X
      XY(2,I)=Y
      WRITE(3,104)I,X,Y
      GO TO 2
   20 WRITE(3,106)NDF
      RETURN
  100 FORMAT('0NODE NO.   X-COORD      Y-COORD')
  102 FORMAT(I5,2F10.5)
  104 FORMAT(I6,2X,2E12.4)
  106 FORMAT( '0SYSTEM HAS',I4,' DEGREES OF FREEDOM')
      END


      SUBROUTINE PF01B(IE,PR,P,G,PF)
C  CONSISTENT LOAD VECTOR FOR LINERALY VARYING PRESSURE ON AN
C     18 DOF TRIANGLE           W H BOWES    JAN 25 1972
      DIMENSION PR(3,1),P(7,7),PF(18),PP(21),G(21,18)
      PC=PR(1,IE)
      PX=PR(2,IE)
      PY=PR(3,IE)
      I=0
      DO 2  K=1,6
      DO 2 JJ=1,K
      II=K+1-JJ
      I=I+1
    2 PP(I)=PC*P(II,JJ)+PX*P(II+1,JJ)+PY*P(II,JJ+1)
      DO 6 I=1,18
      C=0.
      DO 4 J=1,21
    4 C=C+G(J,I)*PP(J)
    6 PF(I)=C
      RETURN
      END


      SUBROUTINE RU01B(U,TRIG,NDF)
C  RESTORES U-COMPONENTS TO X AND Y DIRECTIONS   W H BOWES  AUG 30 1971
      DIMENSION U(1),TRIG(1)
      DO 20 I=2,NDF,2
      IF(TRIG(I))5,20,5
    5 I1=I-1
      US=U(I1)
      U(I1)=TRIG(I1)*U(I1)-TRIG(I)*U(I)
      U(I)=TRIG(I)*US+TRIG(I1)*U(I)
```

```
   20 CONTINUE
      RETURN
      END
      SUBROUTINE RU02B(U,NU,T,NF)
C  RESTORES U-COMPONENTS AT NODE N TO GLOBAL DIRECTION    W H BOWES APR 30 1973
      DIMENSION U(1),T(NF,NF),UP(10)
      DO 2 I=1,NF
      IU=NU+I
    2 UP(I)=U(IU)
      DO 6 I=1,NF
      SUM=0.
      DO 4 IJ=1,NF
    4 SUM=SUM+T(I,IJ)*UP(IJ)
      IU=NU+I
    6 U(IU)=SUM
      RETURN
      END

      SUBROUTINE SS01B(D,B,C,AM,UE,T,S)
C  CALCULATES STRESSES IN SHELL ELEMENTS    W H BOWES  AUG 30 1971
      DIMENSION D(4,4),B(4,6),C(6,6),AM(6,6),UE(1),S(1),SS(6),SSS(6)
      DO 3 I=1,6
      A=0.
      DO 2 IJ=1,6
    2 A=A+AM(I,IJ)*UE(IJ)
    3 SSS(I)=A
      DO 5 I=1,6
      A=0.
      DO 4 IJ=1,6
    4 A=A+C(I,IJ)*SSS(IJ)
    5 SS(I)=A
      DO 7 I=1,4
      A=0.
      DO 6 IJ=1,6
    6 A=A+B(I,IJ)*SS(IJ)
    7 SSS(I)=A
      DO 9 I=1,4
      A=0.
      DO 8 IJ=1,4
    8 A=A+D(I,IJ)*SSS(IJ)
    9 SS(I)=A
      DO 10 I=1,2
      S(I)=SS(I)/T
   10 S(I+2)=SS(I+2)*6./T**2
      RETURN
      END

      SUBROUTINE ST03B(D,B,AI,ET,U,NCON,IE)
C  STRESSES IN CONSTANT STRAIN TRIANGLE DUE TO DISPLACEMENT AND
C     TEMPERATURE CHANGE    W H BOWES  AUG 30 1971
      DIMENSION D(3,3),B(3,6),AI(6,6),U(1),NCON(3,1),UE(6),STR(3),P(6),E
     1P(3)
      DO 2 K=1,2
      DO 2 IH=1,3
      I=3*(K-1)+IH
      II=2*NCON(IH,IE)+K-2
    2 UE(I)=U(II)
      DO 4 I=1,6
      P(I)=0.
      DO 4 IJ=1,6
    4 P(I)=P(I)+AI(I,IJ)*UE(IJ)
      EP(1)=-ET
      EP(2)=-ET
      EP(3)=0.
      DO 6 I=1,3
      DO 6 IJ=1,6
    6 EP(I)=EP(I)+B(I,IJ)*P(IJ)
      DO 8 I=1,3
      STR(I)=0.
      DO 8 IJ=1,3
```

```
    8 STR(I)=STR(I)+D(I,IJ)*EP(IJ)
      A=.5*(STR(1)+STR(2))
      C=STR(1)-A
      R=SQRT(C**2+STR(3)**2)
      IF(C)9,10,10
    9 R=-R
   10 S1=A+R
      S2=A-R
      TH=90./3.14159*ATAN(STR(3)/C)
      WRITE(3,100)IE,(STR(I),I=1,3),TH,S1,S2
      RETURN
  100 FORMAT(I6,2X,3E11.3,F10.1,2E10.3)
      END


      SUBROUTINE ST06B(DB,AI,U,NCON,IE)
C     STRESSES IN A  L S T          W H BOWES    SEPT 11 1971
      DIMENSION DB(3,12),AI(12,12),U(1),NCON(6,1),UE(12),STR(3),P(12)
      DO 2 IA=1,2
      DO 2 IB=1,6
      I=6*(IA-1)+IB
      II=2*NCON(IB,IE)+IA-2
    2 UE(I)=U(II)
      DO 4 I=1,12
      P(I)=0.
      DO 4 IJ=1,12
    4 P(I)=P(I)+AI(I,IJ)*UE(IJ)
      DO 6 I=1,3
      STR(I)=0.
      DO 6 IJ=1,12
    6 STR(I)=STR(I)+DB(I,IJ)*P(IJ)
      A=.5*(STR(1)+STR(2))
      C=STR(1)-A
      R=SQRT(C**2+STR(3)**2)
      IF(C)9,10,10
    9 R=-R
   10 S1=A+R
      S2=A-R
      TH=90./3.14159*ATAN(STR(3)/C)
      WRITE(3,100)IE,(STR(I),I=1,3),TH,S1,S2
      RETURN
  100 FORMAT(I6,3X,3E10.3,F9.1,2E10.3)
      END


      SUBROUTINE ST07B(XY,NCON,NE,D,U)
C     STRESSES AT CENTER OF QUAD RING ELEMENT   W H BOWES   AUG 30 1971
      DIMENSION XY(2,1),NCON(4,1),D(4,4),U(1)
      DIMENSION XYE(4,2),B(4,8),FJ(2,2),EX(2,4)
      DIMENSION DB(4,8),UE(8),STR(4)
      WRITE(3,100)
      EX(1,1)=-1.
      EX(1,2)=1.
      EX(1,3)=1.
      EX(1,4)=-1.
      EX(2,1)=-1.
      EX(2,2)=-1.
      EX(2,3)=1.
      EX(2,4)=1.
      DO 1 I=1,4
      DO 1 J=1,8
    1 B(I,J)=0.
      DO 40 IE=1,NE
      DO 2 I=1,4
      IJ=NCON(I,IE)
      UE(I)=U(2*IJ-1)
      UE(I+4)=U(2*IJ)
      XYE(I,1)=XY(1,IJ)
    2 XYE(I,2)=XY(2,IJ)
      Y=.25*(XYE(1,2)+XYE(2,2)+XYE(3,2)+XYE(4,2))
      RY=1./Y
      DO 13 I=1,2
```

```
      DO 13 J=1,2
      FJIJ=0.
      DO 12 IJ=1,4
   12 FJIJ=FJIJ+.25*EX(I,IJ)*XYE(IJ,J)
   13 FJ(I,J)=FJIJ
      DET=FJ(1,1)*FJ(2,2)-FJ(2,1)*FJ(1,2)
      RDET=1./DET
      SJ=FJ(1,1)
      FJ(1,1)=FJ(2,2)*RDET
      FJ(2,2)=SJ*RDET
      FJ(2,1)=-FJ(2,1)*RDET
      FJ(1,2)=-FJ(1,2)*RDET
      DO 20 J=1,4
      JJ=J+4
      B(3,JJ)=.25*RY
      B(1,J)=.25*(FJ(1,1)*EX(1,J)+FJ(1,2)*EX(2,J))
      B(4,J)=.25*(FJ(2,1)*EX(1,J)+FJ(2,2)*EX(2,J))
      B(2,JJ)=.25*(FJ(2,1)*EX(1,J)+FJ(2,2)*EX(2,J))
   20 B(4,JJ)=.25*(FJ(1,1)*EX(1,J)+FJ(1,2)*EX(2,J))
      DO 25 I=1,4
      DO 25 J=1,8
      DBIJ=0.
      DO 24 IJ=1,4
   24 DBIJ=DBIJ+D(I,IJ)*B(IJ,J)
   25 DB(I,J)=DBIJ
      DO 26 I=1,4
      STR(I)=0.
      DO 26 IJ=1,8
   26 STR(I)=STR(I)+DB(I,IJ)*UE(IJ)
      A=.5*(STR(1)+STR(2))
      H=STR(1)-A
      R=SQRT(H**2+STR(4)**2)
      IF(H)30,31,31
   30 R=-R
   31 S1=A+R
      S2=A-R
      TH=90./3.14159*ATAN(STR(4)/H)
      WRITE(3,102)IE,STR(3),TH,S1,S2
   40 CONTINUE
      RETURN
  100 FORMAT('0ELEM.NO. CIRC.STR. THETA      STR1         STR2')
  102 FORMAT(I6,E13.4,F6.1,E12.4,E12.4)
      END


      SUBROUTINE ST10B(NB,NCB,XY,U,YMB)
C   STRESS IN BAR ELEMENTS          W H BOWES     OCT 14 1971
      DIMENSION NCB(2,1),XY(2,1),U(1)
      WRITE(3,100)
      IF(NB)6,6,2
    2 DO 4 IB=1,NB
      NN1=NCB(1,IB)
      NN2=NCB(2,IB)
      X=XY(1,NN2)-XY(1,NN1)
      Y=XY(2,NN2)-XY(2,NN1)
      EL=YMB/(X**2+Y**2)
      STR=EL*(X*(U(2*NN2-1)-U(2*NN1-1))+Y*(U(2*NN2)-U(2*NN1)))
    4 WRITE(3,102)IB,STR
    6 RETURN
  100 FORMAT('0BAR NO.   STRESS PSI')
  102 FORMAT(I6,3X,E12.4)
      END


      SUBROUTINE ST19B(D,U,NCON,TH,NN)
C   DETERMINES MOMENTS AND STRESSES AT CENTROID OF 18-DOF
C     TRIANGLE IN BENDING               W H BOWES  JUNE 6 1973
      DIMENSION U(1),D(3,3),EM(3),STR(3),C(3)
      DO 20 IN=1,NN
      C(1)=U(6*IN-2)
      C(2)=U(6*IN)
      C(3)=2.*U(6*IN-1)
```

```
      DO 8 I=1,3
      S=0.
      DO 6 IJ=1,3
    6 S=S+D(I,IJ)*C(IJ)
      EM(I)=S
    8 STR(I)=-6.*S/TH**2
      A=.5*(STR(1)+STR(2))
      B=STR(1)-A
      R=SQRT(B**2+STR(3)**2)
      IF(B)9,10,10
    9 R=-R
   10 S1=A+R
      S2=A-R
      THE=90./3.14159*ATAN(STR(3)/B)
   20 WRITE(3,100)IV,EM,STR,THE,S1,S2
      RETURN
  100 FORMAT(I6,3X,6E10.3,F9.1,2E10.3)
      END


      SUBROUTINE TI01B(NP,X,Y,P)
C   INTEGRATES POLYNOMIAL TERMS OVER TRIANGLE    W H BOWES  JAN21 1972
      DIMENSION X(4),Y(4),P(NP,NP)
      NL=NP-1
      X1=X(1)
      X2=X(2)
      X3=X(3)
      Y1=Y(1)
      Y2=Y(2)
      Y3=Y(3)
      A=0
      DO 1 I=1,3
    1 A=A+.5*(X(I)*Y(I+1)-X(I+1)*Y(I))
      P(1,1)=1.
      P(1,2)=0.
      P(2,1)=0.
      DO 14 N=2,6
      IF(N-NL)3,3,20
    3 K=N-1
      GO TO (4,5,5,7,8),K
    4 C=1./12.
      GO TO 10
    5 C=1./30.
      GO TO 10
    7 C=2./105.
      GO TO 10
    8 C=1.
   10 NE=N+1
      DO 12 I=2,N
      I1=I-1
      J1=NE-I
      J=J1+1
   12 P(I,J)=C*(X1**I1*Y1**J1+X2**I1*Y2**J1+X3**I1*Y3**J1)
      P(1,NE)=C*(Y1**N+Y2**N+Y3**N)
      P(NE,1)=C*(X1**N+X2**N+X3**N)
   14 CONTINUE
      P(7,1)=P(7,1)/84.+15./14.*P(5,1)*P(3,1)
      P(6,2)=-P(6,2)/168.+15./14.*(2.*P(5,1)*P(2,2)+5.*P(4,1)*P(3,2)+2.*
     1P(3,1)*P(4,2))
      P(5,3)=-11.*P(5,3)/840.+1./14.*(6.*P(5,1)*P(1,3)+24.*P(3,1)*P(3,3)
     1+30.*P(4,1)*P(2,3)+48.*P(4,2)*P(2,2)+75.*P(3,2)**2)
      P(4,4)=-P(4,4)/56.+1./28.*(25.*P(4,1)*P(1,4)+225.*P(3,2)*P(2,3)+36
     1.*P(3,1)*P(2,4)+36.*P(1,3)*P(4,2)+108.*P(2,2)*P(3,3))
      P(3,5)=-11.*P(3,5)/840.+1./14.*(6.*P(1,5)*P(3,1)+24.*P(1,3)*P(3,3)
     1+30.*P(1,4)*P(3,2)+48.*P(2,4)*P(2,2)+75.*P(2,3)**2)
      P(2,6)=-P(2,6)/168.+15./14.*(2.*P(1,5)*P(2,2)+5.*P(1,4)*P(2,3)+2.*
     1P(1,3)*P(2,4))
      P(1,7)=P(1,7)/84.+15./14.*P(1,5)*P(1,3)
   20 DO 24 I=1,NP
      JL=NP+1-I
      DO 24 J=1,JL
```

```
   24 P(I,J)=A*P(I,J)
      RETURN
      END


      SUBROUTINE TM02B(TH,T)
C     TRANSFORMATION FOR ROTATED AXES,  PLATE IN BENDING   W H BOWES  JUNE 13 1973
      DIMENSION T(5,5)
      TH=3.14159*TH/180.
      C=COS(TH)
      S=SIN(TH)
      CC=C*C
      CS=C*S
      SS=S*S
      T(1,1)=C
      T(1,2)=-S
      T(2,1)=S
      T(2,2)=C
      T(3,3)=CC
      T(3,4)=-2.*CS
      T(3,5)=SS
      T(4,3)=CS
      T(4,4)=CC-SS
      T(4,5)=-CS
      T(5,3)=SS
      T(5,4)=2.*CS
      T(5,5)=CC
      RETURN
      END
```

References

[1] Bell, K., "A Refined Triangular Plate Bending Element," *International Journal of Numerical Methods in Engineering* 1 (1969): 101-122.

[2] Clough, R. W., "The Finite Element Method in Structural Mechanics," *Stress Analysis,* edited by O.C. Zienkiewicz, and G. S. Holister, (London: Wiley, 1965).

[3] Desai, C. S., and J. F. Abel, *Introduction to the Finite Element Method* (New York: Van Nostrand Reinhold, 1972).

[4] Flügge, W., *Stresses in Shells,* 2nd ed. (New York: Springer-Verlog, 1973).

[5] Holand, I., and K. Bell, *Finite Element Methods in Stress Analysis* (Trondheim, Norway: Tapir, The Technical University of Norway, 1969).

[6] Hrennikoff, A., "Solutions in Problems of Elasticity by the Framework Method," *Journal Applied Mechanics* 8 (4) (December 1941): A169-A175.

[7] James, M. L., G. M. Smith, and T. C. Wolford, *Applied Numerical Methods*, (Scranton: International Textbook, 1967).

[8] Kuo, S. S., *Numerical Methods and Computers* (Don Mills, Ontario: Addison-Wesley, 1966).

[9] Megård, G., "Analysis of Thin Shells Using Planar and Curved Finite Elements," Report to be published at the Division of Structural Mechanics, The Technical University of Norway, Trondheim, Norway.

[10] Payne and Irons, referred to as private communication in *The Finite Element Method* by Zienkiewicz.

[11] Rosanoff, R. A. and T. A. Ginsberg, "Matrix Error Analysis for Engineers," AFFDR-TR-66-80.

[12] Timoshenko, S. P., and J. N. Goodier, *Theory of Elasticity,* 3rd ed. (New York: McGraw-Hill, 1970).

[13] Timoshenko, S. P , and S. Woinowsky-Krieger, *Theory of Plates and Shells,* 2nd ed. (New York: McGraw-Hill, 1959).

[14] Turner, M. J.; R. W. Clough; H. C. Martin; and L. J. Topp, "Stiffness and Deflection Analysis of Complex Structures," *Journal of the Aeronautical Sciences* 23 (9) (September 1956): 805-823.

[15] Williams, D., "A General Method (depending on the aid of a digital computer) for Deriving the Structural Influence Coefficients of Aero-

plane Wings," Report No. Structures 168, Royal Aircraft Establishment, Farnborough, U.K., November 1954.

[16] Zienkiewicz, O. C., *The Finite Element Method in Engineering Science*, 2nd edition (New York: McGraw-Hill, 1971).

Index

Index

About the Authors

William H. Bowes studied at Dalhousie University, the Nova Scotia Technical College and the University of Michigan and received degrees in Mechanical Engineering, Civil Engineering and Engineering Mechanics. He has taught in a variety of fields at Dalhousie and at Carleton University where he is now professor of engineering. He was Stress Analyst in the design team that pioneered the CANDU type of nuclear power reactor. In recent years Professor Bowes has been active in promoting the Finite Element Method as a design tool and, with Dr. Russell, has offered several short courses for practicing engineers. To meet the need for a textbook illustrating the Finite Element Method as a design tool, the authors have written *Stress Snalysis by the Finite Element Method for Practicing Engineers*.

Leslie T. Russell received the B.Eng. from the Nova Scotia Technical College, the M.Sc. from Queen's University and the Ph.D. from Carleton University. He has taught at the Nova Scotia Agricultural College, Dalhousie University and Carleton University, and is an associate professor of Mechanical Engineering at the Nova Scotia Technical College. Dr. Russell has been associated with the Defence Research Board of Canada as a member of the design team for the Variable Depth Sonar Equipment now in use by many countries. In recent years he has been active in association with Professor Bowes in promoting the use by practicing engineers of the Finite Element Method.